With the Compliments of
H.R.H. The Prince of Wales
and
The Prince of Wales Foundation

Harmony

A NEW WAY OF
LOOKING AT OUR WORLD

Harmony

A NEW WAY OF
LOOKING AT OUR WORLD

HRH
THE PRINCE
OF WALES

with

TONY JUNIPER
IAN SKELLY

HARPER

An Imprint of HarperCollins*Publishers*
www.harpercollins.com

HarperCollins books may be purchased for educational, business, or sales
promotional use. For information, please write: Special Markets Department,
HarperCollins Publishers, 10 East 53rd Street, New York, NY 10022.

Published in Great Britain in 2010 by Blue Door,
an imprint of HarperCollins Publishers.

FIRST EDITION

Designed by Birdy Book Design

Library of Congress Cataloging-in-Publication Data has been applied for.

ISBN: 978-0-06-173131-0

10   11   12   13   14   HCUK   10   9   8   7   6   5   4   3   2   1

Harmony 1

Finds tongues in trees,
books in the running brooks,
Sermons in stones,
and good in every thing.

WILLIAM SHAKESPEARE

This is a call to revolution. The Earth is under threat. It cannot cope with all that we demand of it. It is losing its balance and we humans are causing this to happen.

'Revolution' is a strong word and I use it deliberately. The many environmental and social problems that now loom large on our horizon cannot be solved by carrying on with the very approach that has caused them. If we want to hand on to our children and grandchildren a much more durable way of operating in the world, then we have to embark on what I can only describe as a 'Sustainability Revolution' – and with some urgency. This will involve our taking all sorts of dramatic steps to change the way we consider the world and act in it, but I believe we have the capacity to take these steps. All we have to see is that the solutions are close at hand.

The Earth's alarm bells are now ringing loudly and so we cannot go on endlessly prevaricating by finding one sceptical excuse after another for avoiding the need for the human race to act in a more environmentally benign way – which really means only one thing: putting Nature back at the heart of our considerations once more. But that is only the start of it. We must go much further. 'Right action' cannot happen without 'right thinking' and in that simple truth lies the deeper purpose of this book.

For more than thirty years I have been working to identify the best solutions to the array of deeply entrenched problems we now face. I have tried to do this, for instance, by demonstrating the principles of what I believe to be truly 'sustainable' agriculture through organic farming. I have tried to demonstrate the principles of 'sustainable' urbanism which can add social and environmental value to towns and cityscapes through mixed-use development, by placing the pedestrian at the centre of the design process, by emphasizing local identity and character and by the use of ecological building techniques. For many years I

PREVIOUS SPREAD: *The Earth seen from Apollo 8 as it orbited the Moon on 24th December 1968. As far as we know Earth is the only place in the Universe that harbours any life. This image is credited with inspiring the modern environmental movement.*

LEFT: *'As our planet's life-support system begins to fail and our very survival as a species is brought into question, remember that our children and grandchildren will ask not what our generation said, but what it did. Let us give an answer, then, of which we can be proud.' Part of my keynote address to world leaders at the UN Conference on Climate Change in Copenhagen, December 15, 2009.*

have been working to create effective partnerships between the private, public and non-governmental organization (NGO) sectors, not only to address the serious threats posed by climate change, but also to create major initiatives to try to save what is left of the world's rainforests, as well as other major natural ecosystems – such as oceans and wetlands – which are now under dire threat of collapse. I have also tried for twenty-five years to encourage social and environmentally responsible business; to suggest a more balanced approach to certain aspects of medicine and healthcare; more rounded ways of educating our children and a more benign, 'whole-istic' approach to science and technology. The trouble is that in all these areas I have been challenging the

accepted wisdom; the current orthodoxy and conventional way of thinking, much of it stemming from the 1960s but with its origins going back over 200 years.

Perhaps I should not have been surprised that so many people failed to fathom what I was doing. So many appeared to think – or were told – that I was merely leaping from one subject to another – from architecture one minute to agriculture the next – as if I spent a morning saving the rainforests, then in the afternoon jumping to help young people start new businesses.

What I have actually been trying to demonstrate is that all of these subjects are completely inter-related and that we have to look at the whole picture to understand the problems we face. For not only does it concern the way we treat the world around us, it is also to do with how we view ourselves.

In all my efforts I have tried to make it clear that all these subjects suffer the same problems because they have become detached from important basic principles – the principles that produce the active state of balance which is just as vital to the health of the natural world as it is for human society. We call this active but balanced state 'harmony' and this book is dedicated to explaining how harmony works.

It is a book in which I hope to share the results of much thought, observation and reflection over the past thirty or forty years. I want to show what I have gained and achieved from studying the essential principles of harmony – how they work in Nature and how, if we ignore or flout them, the Earth's precious life-support systems start to wobble and eventually may collapse. In some cases they have already fallen into a perilous state.

That is why our journey begins with a look at just what we are doing to our life-giving Earth after some two and a half centuries of intensive industrialization. We all hope for solutions and that is why I want to end this journey by offering what might turn out to be a few of them, but the solutions must be understood in their proper context. I know from experience that if any solution is not deeply rooted in the right principles it will be of no use in the long term. In fact, quite the contrary, it will tend to compound the problems we already have. That is why I also want to put our present situation in its true historical context. We have to realize that we are travelling on the wrong road, but we need to understand why.

It is very strange that we carry on behaving as we do. If we were on a walk in a forest and found ourselves on the wrong path, then the last thing we would do is carry on walking in the wrong direction. We would instead retrace our steps, go back to where we took the wrong turn, and follow the right path. This is why I feel it is so important to offer not just an overview of our present situation and not just a list of the solutions. I certainly want the world to wake up to the fact that we are travelling in a very dangerous direction, but it is crucial that we retrace how this has come to be, otherwise we will not head onto a better path in the future.

Crisis of perception

I would suggest that one of the major problems that increasingly confronts us is that the predominant mode of thinking keeps us firmly on this wrong path. When people talk of things like an 'environmental crisis' or a 'financial crisis' what they are actually describing are the consequences of a much deeper problem which comes down to what I would call a 'crisis of perception'. It is the way we see the world that is ultimately at fault. If we simply concentrate on fixing the outward problems without paying attention to this central, inner problem, then the deeper problem remains, and we will carry on casting around in the wilderness for the right path without a proper sense of where we took the wrong turning.

That is why I wanted to put this book together. With Tony Juniper and Ian Skelly's help, I want to demonstrate that we have grown used to looking at the world in a particular way that obscures the danger of a very disconnected approach. All of the solutions I want to suggest depend for their success upon looking at the world in a different way. It is not strictly a new way and that is why we will travel back in time to see the world as the ancients saw it, but it is a way of seeing things that stands very much at odds with what has become the only reasonable way of looking at the world. If that reaction starts to grow then I urge you to hold onto one important fact, that this timeless view of things is rooted in the human condition and in human experience.

It may be a bit daunting if I suggest at the outset that I want to include in this journey a brief tour of 'traditional philosophy' but I can assure you that such an explanation will be painless and that everything will be explained simply. Not least because it *is* simple.

Perhaps it is worth remembering what that word 'philosophy' means. It is a combination of two Greek words: one meaning 'love of' and the other meaning 'wisdom'. So, to be a 'philosopher' is to be a lover of wisdom, and the wisdom this refers to is human wisdom, of the sort that has been handed down from generation to generation in all societies throughout the world. Until quite recently, this time-honoured wisdom framed the way all civilizations behaved. It emphasized the right way to see our relationship with the natural world, it taught in practical ways how to work with the grain of Nature rather than

against it, and it warned of the dangers of overstepping the limits imposed by Nature *on herself*. In short, this wisdom emphasized the need for, and the means of maintaining, harmony.

Ancient wisdom

As I first struggled to understand what age-old thinking like this could teach us, I began to notice a curious connection between the many problems our modern world view had created and a subject that increasingly fascinated me.

The five-petalled rose pattern traced over 8 years in the skies above Earth by our nearest neighbour, Venus, depicted 400 years ago by the German astronomer, Johannes Kepler. It is the source of the familiar five-pointed star found in many natural forms and in the world's sacred architecture.

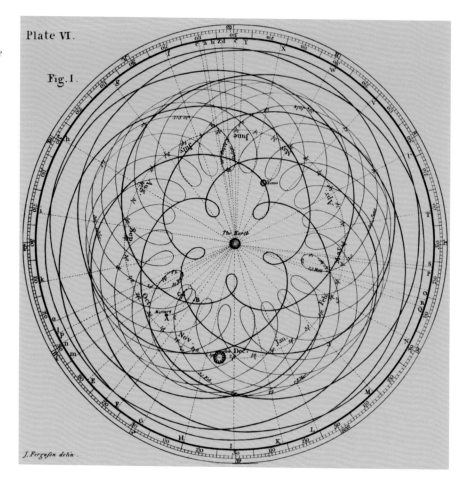

Plate **VI**.

Fig. I.

J. Fergufon delin.

It was a surprising subject. It was the design and symbolism of the architecture of the temples, mosques, and cathedrals of the world. The more I learned about it, the more I became aware that there was a similarity between the way ancient civilizations built their sacred structures and the way the natural world itself is structured and behaves. The ratios and proportions that define the way natural organisms grow and unfold are the same as those that underpin the structure of the most famous ancient buildings. I was among a number of people who began to piece together a great jigsaw which revealed, much to my surprise, a profound insight into what really lay at the heart of ancient thinking. I shall explain this with lots of images in the section called 'The Grammar of Harmony' in Chapter 3, which gives context to the history of modernity, simply because there is a direct relationship between the patterns that inspired the builders of all those great masterpieces of sacred architecture and the way the natural world operates when it is in a healthy state of balance. The two speak with the same 'grammar'.

Seeing this, I began to realize that the great juggernaut of industrialization relies upon a somewhat aberrant kind of language – a man-made one – which articulates a world view that ignores Nature's grammar. Much of the syntax of this synthetic language is out of synchrony with Nature's patterns and proportions and this is why it so often jars with the language of Nature. This is why so many Modernist buildings don't feel 'right' to so many people, even though they may find them clever; or perhaps why we feel uncomfortable with factory farming, even though it makes economic sense because it supplies such a lot of food at such low prices; or why we feel something is missing from a form of medicine that treats the body like a machine and does not accommodate the needs of the mind or the spirit.

I find, by contrast, that if people are encouraged to immerse themselves in Nature's grammar and geometry – discovering how it works, how it controls life on Earth, and how humanity has expressed it in so many great works of art and architecture – they are often led to acquire some remarkably deep philosophical insights into the meaning and purpose of Nature and into what it means to be aware and alive in this extraordinary Universe. This is particularly so in young people and the results of such immersion are as heartening as they are surprising. They help to point to the changes in thinking that we need to make to achieve the wider vision of a Sustainability Revolution.

Essentially it is the spiritual dimension to our existence that has been dangerously neglected during the modern era – the dimension which is related to our intuitive feelings about things. The increasing tendency in mainstream Western thinking to ignore this spiritual dimension comes from a combination of the growth in cynicism during the latter half of the twentieth century and the wholesale dismissal of the big philosophical questions about our existence. The dominant world view only accepts as fact what it sees in material terms and this opens us up to a very dangerous state of affairs, not least because the more extreme this approach becomes, the more extreme the reaction tends to be at the other end of the scale, so we end up with two fundamentalist, reductionist camps that oppose each other. On the one side, a fundamentalist secularism and on the other, fundamentalist religions. This seems to happen in Christianity as it does in Islam and, wherever it happens, the more puritanical and literal the religious interpretation becomes, the more a culture abandons and then even attacks the age-old, symbolic interpretations of its own tradition – those teachings which actually emphasize the necessary limits to our behaviour. With so much emphasis on the historical accuracy of the origins of a religion, the search for mystery appears to give way to a vain search for certainty. What was a traditional attitude becomes a 'modern' and 'progressive' one, all too intolerant of restraint, and so the limits – Nature's necessary limits – end up being overshadowed by dogma.

Has this come about as a result of one dimension of our outlook becoming too dominant in our thinking? And if so, what is the nature of that outlook? Having considered these questions long and hard, my view is that our outlook in the Westernized world has become far too firmly framed by a mechanistic approach to science, the one that has increasingly prevailed in the West for the past four hundred years. This approach is entirely based upon the gathering of the results that come from subjecting physical phenomena to scientific experiment. It is called 'empiricism' and it is, if you like, a kind of language. It is a very fine one, but it is a language not able to fathom experiences like faith and the meaning of things. Nor can it articulate matters of the soul. It is now the only popularly trusted level of language we may use to articulate our understanding of the world. Don't get me wrong, it has a very valuable role to play, but the trouble is, empiricism now assumes authority beyond the area it is capable of considering and, consequently, it excludes the voices of those other levels of language that once played their rightful part in giving humanity a comprehensive view of reality – that is, the philosophical and the spiritual levels of language. This is why it conveniently elbows the soul out of the picture.

Think of something as basic as the conversation that might take place in a biology lesson where a science teacher is called upon by pupils to address the moral and ethical questions of whether or not it is a good thing to manipulate genes. At that point, does the teacher act as a philosopher or remain a science teacher? I am pretty sure that the majority of teachers would certainly feel very uncomfortable about assuming the role of spiritual guide when such questions arise. The essential point here is, how far our empirical knowledge can go before it begins to encroach on territory it is not qualified to discuss. Let me be clear about it. Science can tell us how things work, but it is not equipped to tell us what they mean. That is the domain of philosophy and religion and spirituality.

Let me say again – empiricism has its part to play, but it cannot play all of the parts. And yet, because it tries to, we end up with the general outlook that now prevails. The language of empiricism is now so much in the ascendant that it has authority over any other way of looking at the world. *It* decides whether those other ways of looking at things stand up to *its* tests and therefore whether they are right or wrong.

This has not always been the case. A specifically mechanistic science has only recently assumed a position of such authority in the world and I want to show how this came to be: how its influence from the seventeenth century onwards spread, and slowly but surely excluded those other levels of language that were once much more a part of the conversation. For not only has it prevented us from considering the world philosophically any more, our predominantly mechanistic way of looking at the world has also excluded our spiritual relationship with Nature. Any such concerns get short shrift in the mainstream debate about what we do to the Earth. They are dismissed as outdated and irrelevant because a thing does not exist if it cannot be weighed or measured. And so we live in an age which claims not to believe in the soul. Empiricism has proved to us how the world really fits together and how it really works and, on its terms, this has nothing to do with God. There is no empirical evidence for the existence of God, so therefore God does not exist. That seems a very reasonable, rational argument, so long as you go along with the empirical definition of God as a 'thing'. I presume the same argument can also be

Behind the familiar images of sacred sites like this figure of Christ on Canterbury Cathedral in England, is a symbolism that goes beyond the particular culture and time in which it was created.

applied to the existence of thought. After all, no brain-scanner has ever managed to photograph a thought, nor a piece of love for that matter, and it never will, so, by the same terms, thought and love do not exist either.

That may appear flippant, but my point is that this is the consequence of doggedly following Galileo's line that there is nothing in Nature but quantity and motion. Over time it has added up to a serious situation where we are no longer able to view the world much beyond its surface and its appearance. We are persuaded, instead, to follow a way of being that denies the non-material side to our humanity even though, contrary to what is supposed to be a growing popular belief, this other half of ourselves is actually just as important as our rational side, if not more so. It is our means of relating to the rest of the natural world and this is why I have long felt so alarmed that our collective thinking and predominant way of doing things are so dangerously out of balance with Nature. We have come to function with a one-sided, materialistic approach

that is defined not by its inclusiveness, but by its dismissal of those things that cannot be measured in material terms.

This is peculiar to the history of the West. In general, people from elsewhere in the world do not understand how Nature has become so secularized. Even many people in the West fail to recognize that so much modern science is not simply an 'objective' knowledge of Nature, but is based upon a particular way of thinking about existence and geared to the ambition to gain dominion over Nature. The way in which this has happened has a lot to do with the numbing of our vital inborn or 'inner tutor', the so-called human 'intuition'.

Our intuition is deeply rooted in the natural order. It is 'the sacred gift', as Einstein called it. Many sacred traditions refer to it as the voice of the soul: the link between the body and mind and therefore the link between the particular and the universal. If we were to recognize this, we would perhaps once again begin to see our existence in its proper place within creation and not in some specially protected and privileged category of our own making. That is hardly likely to happen as long as scientific rationalism continues to turn people away from any form of spiritual practice or reflection by perpetuating what seems to me to be a widespread confusion. It often comes to light during one of those typical interrogations of a person who experiences faith. They are expected to give empirical proof that God exists. As I hope will become clear later, this question can only be taken seriously when faith and the Divine are regarded as material objects.

A much more integrated view of the world and our relationship with it existed throughout ancient history and right up to that critical period in seventeenth-century Europe when Western thinking began to take a more fragmented view of things. It is not so much the fragmentation, but its causes that I have come to see are the linchpins of the problem and that is why I feel

At work at Highgrove, my home in Gloucestershire, England, laying hedges using age-old traditional techniques. Hedgerows are not only long-lasting, sturdy ways of keeping stock in fields, they are havens for wildlife and are a time-honoured way of stopping the erosion of top soil.

it necessary to explore, in the lightest way possible, how the modern world was born and how we came to regard the world in the overtly 'mechanistic' way we do today. By persisting in this view, we ignore, abandon and waste the wisdom, knowledge and skills that have been built up over the entire course of human history. It is, perhaps, not so understood as it should be that so much of the wisdom I am referring to came to humanity from revelation. Revelation is not deemed possible from an empirical point of view. It comes about when a person practises great humility and achieves a mastery over the ego so that 'the knower and the known' effectively become one. And from this union flows an understanding of 'the mind of God'. I cannot stress it firmly enough: by dismissing such a process and discarding what it offers to humankind, we throw away a lifebelt for the future.

If people are encouraged to immerse themselves in Nature's grammar and geometry they are often led to acquire some remarkably deep philosophical insights.

I was born in 1948, right in the middle of the twentieth century, which had dawned amid the gleaming Age of the Machine, the very engine of colossal change in the Western world. By the 1920s the overriding desire in every leading nation was for the new and the modern: perhaps a natural reaction as people struggled to recover amid the debris of a shattered world after the Great War. The same thing happened in the wake of the unimaginable horrors of the Second World War as, once again, industrialized nations had to find their feet, and quickly. Such was the sense of a fresh start that, by the mid-1950s, a frenzy of change was sweeping the world in a wave of post-war Modernism, and that created a new age of radical experimentation in every major field of human endeavour. By the 1960s the industrialized countries were well on their way to creating what many imagined would be a limitless Age of Convenience. For those who found themselves riding the juggernaut, life became more comfortable, less painful, and lasted longer.

I remember that period in the 1960s only too well and even as a teenager I felt deeply disturbed by what seemed to have become a dangerously short-

HARMONY

sighted approach. I could not help feeling that in whichever field these changes were taking hold, with industrialized techniques replacing traditional practices, something very precious was being lost. In many cases it was not so much being lost as wilfully destroyed. I also recall the gleeful, fashionable cries of 'God is dead', perhaps the epitome of this short-sightedness. It certainly offered an early clue as to what had happened to our collective view of the natural world.

Such was the dogma of the day that when eventually, in the 1970s, I began to raise these concerns publicly, I had to face an avalanche of criticism that was nearly all based on a very basic misunderstanding. Most critics imagined that I somehow wanted to turn the clock back to some mythical Golden Age when all was a perfect rural idyll. But nothing could be further from the truth.

My concern from the very start was that Western culture was accelerating away from values and a perspective that had, up until then, been embedded in its traditional roots. The industrialization of life was becoming comprehensive and Nature had become 'secularized'. I could see very clearly that we were growing numb to the sacred presence that all traditional societies still feel very deeply. In the West that sense of the sacred was one of the values that had stood the test of time and had helped to guide countless generations to understand the significance of Nature's processes and to live by her cyclical economy. But, like the children who followed the Pied Piper, it was as if our beguiling machines, not to say four centuries of increasingly being dependent upon a very narrow form of scientific rationalism, had led us along a new but dangerously unknown road – and a dance that has been so merry that we failed to notice how far we were being taken from our rightful home. The net result was that our culture seemed to be paying less and less heed to what had always been understood about the way Nature worked and the limits of her benevolence, and to how, as a consequence, the subtle balance in many areas of human endeavour was being destroyed. What I could see then was that without those traditional 'anchors' our civilization would find itself in an increasingly difficult and exposed position. And, regrettably, that is what has happened.

This is why, ever since those disturbing days, I have expended vast amounts of energy to help save what remains of those traditional approaches. I knew they would be needed for a 'rainy day' which I fear is now close by. However, back then I realized that what mattered was to prove their worth. It was no use arguing about the theory or trying to persuade people that so many of these traditional ways are rooted in a deep-seated ancient, philosophical outlook. That would have to come later, when the world was more sensitive to what had so swiftly been consigned to the shadows. No, the point was to emphasize the principles of harmony that we had lost sight of. I wanted to do this in a

LEFT: *According to UN figures, the US alone buries 222 million tons of household waste a year. China is fast catching up with 148 million tons. As the rubbish degrades it gives off landfill gas, 50% of which is methane and up to 40% is CO_2. Methane is 20 times more effective at trapping heat within the atmosphere, making landfill sites one of the biggest producers of methane gas in the world.*

contemporary way – to find as many ways as possible of reintegrating traditional wisdom with the best of what we can do now so as to demonstrate how we might make this age fit for a sustainable future.

It is probably inevitable that if you challenge the bastions of conventional thinking you will find yourself accused of naivety. And all the more so if you challenge the current world view in all of the important areas of human activity – in agriculture and architecture, education, healthcare, in science, business, and economics. In those early years I was described as old-fashioned, out of touch and anti-science; a dreamer in a modern world that clearly thought itself too sophisticated for 'obsolete' ideas and techniques, but I could see the stakes were already far too high in all of these areas. Even back at the end of the 'swinging sixties' the damage was showing through, and I felt it my duty to warn of the consequences of ignoring Nature's intrinsic tendency towards harmony and balance before it was all too late. What spurred me on was an essential fact of life, an undeniable law: that if we ignore Nature, everything starts to unravel. This is why, from the very beginning, I kept pointing out that it is vital we seek ways of putting Nature back in her rightful place – that is, at the centre of things, and that includes in our imagination as well as in the way we do things.

So what are these timeless 'principles'? Fashions may change, ideologies may come and go, but what remains certain is that Nature works as she has always done, according to principles that we are all familiar with. Nutrients in soils are recycled, rain is generated by forests, and life is sustained by the annual cycles of death and rebirth. Every dead animal becomes food for other organisms. Rotting and decaying twigs and leaves enrich soils and enable plants to grow, while animal waste is processed by microbes and fungi that transform it into yet more vital nutrients. And so Nature replaces and replenishes herself in a completely efficient manner, all without creating great piles of waste.

This entire magical process is achieved through cycles. We all know how the seasons follow one another, but there are many more cycles within those over-arching ones and so many of them are interrelated so that the life cycles of many animals and plants dovetail with one another to keep the bigger cycles moving. For instance, in Spring some songbirds time the hatching of their eggs to coincide with a population explosion in the caterpillars that they feed to their chicks. Built into these many cycles are self-correcting checks and balances whereby the relationships between predators and prey, the rate of tree growth, and the replenishment of soil fertility are all subject to factors that facilitate orderly change and progress through the seasons and keep everything in balance. No single aspect of the natural world runs out of proportion with the others – or at least not for long.

What is more, Nature embraces diversity. The health of each element is enhanced by there being great diversity or, as is now commonly called today, 'biological diversity' or 'biodiversity' for short. The result is a complex web made up of many forms of life. For this web to work best there is a tendency towards variety and away from uniformity and, crucially, no one element can survive for long in isolation. There is a deep mutual interdependence within the system which is active at all levels, sustaining the individual components so that the great diversity of life can flourish within the controlling limits of the whole. In this way, Nature is rooted in wholeness.

There is one other principle or quality I would draw attention to. I will refer to it a lot throughout this book because, in my view, it is extremely important. It is the quality of beauty, which has inspired countless generations of artists and craftsmen. 'Beauty is in the eye of the beholder', it is said, but I have always felt that, because people are as much a part of the whole system of life as every other living thing and because beauty is to be found in the fabric of all that we are, the truth is the other way around. Our ability to see beauty in Nature is entirely consequential on our being a part of Nature Herself. In other words, Nature is the source, not us. In this way, if we do not value beauty then we ignore a vital ingredient in the well-being of the world. This is just as important to recognize as the other elements in my proposition because none of us can survive for very long if the underlying well-being of the planet is destroyed.

I find that the world view which prevails today in Western societies, and in an increasing number of others who are following its flawed logic, pursues priorities that are almost diametrically opposite to those I have just described. There is an emphasis on linear thinking rather than seeing the world in terms of cycles, loops and systems, and the intention is to master Nature and control her, rather than act in partnership. Our ambition is to seek ever more specialized knowledge rather than take a broad or 'whole-istic' view. Nearly all we do generates masses of waste almost as if it is an automatic consequence of how we have to live. Monocultures of crops, of brands, and ideas have come to dominate and crush diversity in our farming, in our culture – and in our business too. Instead of a large number of small actors, we have a small number of huge organizations that now dominate many parts of our economic activity. And, in all we do, we load the atmosphere with those gases that build up a kind of insulating blanket around the Earth, so-called 'greenhouse' gases which accumulate in entirely unnatural quantities which then makes the world ever warmer, thus disturbing the balance that the Earth seeks to maintain. We carry on doing this as if we are immune to the consequences – as if somehow we have

Late Spider Orchid. This rare plant flowers in England during June and July.

isolated ourselves from the inevitable checks that in the end govern all life on Earth.

When I began pointing all this out in those early years when there was not quite so much scientific evidence to back up what my intuition was telling me, it proved a particularly unrewarding occupation. I am relieved to say that now the story is a little different. For one thing, I no longer have to theorize. Now I can point, not only to a vast body of evidence that describes the consequences of our behaviour, but also to an array of successful practical examples of how better to approach matters. These examples, relating to everything from farming to town-planning, are healthier, more beautiful, more human-centred and much more 'sustainable' – although I prefer the word 'durable'. It is these that I plan to explain.

Having also travelled widely in those years and been fortunate to meet and discuss these issues with a large number of people, many of them leading experts in their field, from whose wisdom and knowledge I have benefitted, I also want to share the achievements I have witnessed. We will discover great work being done all over the world, from the UK and the United States to Australia and China, and my hope is that in so many vivid ways it will become clear just what goes wrong if we abandon traditional knowledge and practices and turn away from how Nature behaves.

'Knowledge is power' is a dictum behind much science and experimentation, but is there value in widening the scope of science teaching so that it seeks a deeper understanding of the wholeness of Nature?

The contrast between the way these more harmonious approaches work and the way things are done in the mainstream will, I hope, reveal the many deep cracks in the veneer of our Age of Convenience. These are already becoming more obvious, exposing just how flimsy its foundations really are. We may still enjoy plenty of convenience for now and, of course, it would be marvellous if we could somehow maintain the whole edifice without suffering the eventual consequences of deliberately excluding Nature from the equation in every field known to humankind, but the costs to both the natural world and our own inner world are very severe. We are beginning to recognize the outline of what we have really engineered for ourselves. Not an age of limitless convenience after all, but a much more disturbing 'Age of Disconnection'. That is to say, we have systematically severed ourselves from Nature and the importance to us, as to everything else on Earth, of her processes and cyclical economy. As a result, we are beginning to fall seriously out of joint with the natural order. And there is order. Whether we choose to be part of the process or not, everything in truth depends upon everything else. Whether it is the bee to the flower, the bird to the fruit tree, or the man to the soil, we depend upon them all – and we neglect this simple principle at our peril. It stands to reason: take away the bee and there can be no flower; without the bird there will be less fruit; deplete the soil and very soon people will begin to starve.

Such obvious relationships are taught in these simple terms to small children in primary schools and yet, by the time they reach adulthood, a strange trans-formation appears to have taken place. It is almost as if they have gone through a subtle brainwashing that encourages them to follow the rest of the merry throng and dance without question to the Pied Piper's tune. Like everyone else they become persuaded to think that we *can* do without everything else and that we *can* ignore the essential rhythms and patterns of Nature; that, indeed, nothing is sacred any more, not even that mysterious ordered harmony which ultimately sustains us.

There is little question in my mind now that this is a dangerous course. And that we no longer have a choice. If we could exist independently of Nature and her underlying principles, that would be splendid, but we can't – certainly not if we retain a modicum of interest in our children's and grandchildren's future on this threatened planet. The thought of them has been, for me, the main driving force for this book, regardless of how it may be greeted, and if it moves others to reflection, then let this book be a means of exploring what has caused us to think that we can abandon Nature's rhythmic patterns. We have done so, not just in the mechanized processes we use to grow our food and treat our farm animals, or the way in which we design and build our homes, towns and cities, or the way in which we deny the crucial relationship between mind, body and spirit in healthcare. We have also done so in the way we fail in our systems of economics to measure and put a proper value on Nature's vital services, and even in the manner we teach *out* a proper whole-istic understanding of the fact that we are a *part* of Nature not apart from Her when it comes to our children's education. For they all follow an approach to life that places the greatest value on a mechanistic way of thinking and a linear kind of logic. But carrying on in this way as if, fundamentally, it is 'business as usual' is no longer an option. We cannot solve the problems of the twenty-first century with the world view of the twentieth century.

Terms used in this book

Before we begin our journey there are a number of key words and terms that I will use throughout that I feel I should explain. One of these terms is 'mechanistic thinking'. This stems from what happened in Western thought from the seventeenth century onwards, after the great pioneers of empirical discovery like Descartes and Francis Bacon laid down the principles of the Scientific Revolution. Nature began to be understood in the more clinical terms of its mechanics, as we shall see. This is because, in the main, our science has been based on a 'reductionist' approach. Organisms are broken down and their separate parts are studied in mechanical terms. Hence in schools today children are generally taught to see the human heart as nothing more than a pump, the lungs as a set of bellows, and the brain as some sort of very clever computer with the human mind conveniently explained away as the product of an electromagnetic effect of brain function. Despite the incredible leaps that Quantum Mechanics and Particle Physics and the lessons on the inter-connectivity of matter they so readily offer us, it still appears odd that many people seem not to have a knowledge of these things. Is this, perhaps, why things start to get a bit fuzzy in the schoolroom when it comes to defining consciousness in mechanistic terms or, for that matter, the imagination. Quite where the resonance we feel for the beauty of things or, ultimately, love is anybody's guess. The consequence of this outlook is that we have amassed an extensive database of how the world works that has enabled us to increase the speed and adaptability of many elements of the natural world, but in doing so we have lost a valuable and ancient perspective.

The eighteenth-century agenda of the Enlightenment, based predominantly on the pursuit of progress through science and technology, is so much a part of the furniture today that we do not even question it as an ideology. And yet it

RIGHT: *Animals kept in crates, reared all their lives in sheds and fed on a diet of corn and growth hormones disconnects even the creatures we rely upon for our food from the natural world. Factory farming is said to be the only way to feed the world, but this ignores the massive hidden costs and the need to give back to Nature as much as we take. There are better ways to produce food than this.*

is as if we peer at the world through a letterbox, believing that what our science reveals to us is the whole picture even though science does not itself deal with the meaning of things, nor does it encourage a very joined-up way of working. As a result, time and again one problem is solved, but in its wake many others are created, often far worse than the one we set out to resolve.

To see this in action we only have to consider the way water companies in the UK have to spend around £100 million a year removing pesticides and other chemicals from the water supply. These chemicals are the fallout of a supposedly efficient form of industrialized agriculture – an agriculture that works according to mechanistic thinking. The same mechanistic response is applied in the US, where every year many more millions of dollars are spent blasting fresh meat with ammonia in enormous, gas-guzzling chemical plants to cleanse it of the fatal *E. coli* bug that has blighted the food industry for decades. This bug is only there because of the intensive way in which cattle are reared on a diet of corn on vast 'feed lots' which are, to all intents and purposes, like concentration camps for cattle. Much of the *E. coli* bug could easily be removed from the gut of cattle simply by giving them what they are designed by Nature to eat, which is grass, but that does not automatically follow when mechanistic thinking is at work. The knee-jerk reaction is to use more and very costly technology to solve any problems that arise from the solution to an original problem, and so we spawn yet more problems, each one solved in the same isolated way. Nature has the simpler remedy, but she is excluded from the process. She is no longer involved in the cure.

This fragmented view of the world extends to the way people are expected to behave. I come

across many instances when the absence of this understanding of how we really fit within the great scheme of things forces people to censor what their intuition might be telling them, to the point where I sometimes wonder if there are a considerable number of people living an almost schizophrenic-like existence. The pressure can be enormous on individuals to draw a very clear line between their private feelings and their public, professional occupation. I have lost count of the number of people I have spoken with who tell me quietly of how, even though privately they may feel deeply anxious inside themselves about the consequences of this whole mechanistic approach, when at work they are expected to lock those feelings away and follow the corporate diktat, which so often reflects the mechanistic mindset that can be so destructive of Nature and her systems.

If we continue to engineer the extinction of the last remaining indigenous, traditional societies, we eliminate one of the last remaining sources of that wisdom.

This bizarre denial has far-reaching and serious consequences for the lives of millions of people, and all the more so if it manifests itself in those who effectively run the world. I intend to give graphic details of the ultimate price some of the poorest farmers in India have had to pay because of it. But it is not just the lives of those in developing countries. Many small-scale farmers in the US also find themselves up against the same might of a globalized system that allows only a few giant corporations to control more or less the whole food production and distribution system across an increasing proportion of the world.

I find it revealing that a substantial number of the people who work for such organizations can often feel instinctively anxious about what this current world view expects of them, but they dare not express their disquiet for fear of being considered old-fashioned, not 'on message' or anti-science. They can see quite clearly the long-term implications of what they are being asked to do in their professional lives, but even so, I suspect that if I asked them whether they have any sense of the inner value of things when it comes to the decisions they take, or whether they look beyond the mechanics of Nature to obtain a true sense of what life consists of, the chances are they would feel obliged to accuse me of relying on 'superstition'. They would most certainly fight shy of agreeing that there may be such a thing as an invisible 'pattern' in which all manifestations of life take place. But if they were to realize how many people in the same situation felt the same way about the consequences of what they are doing, I wonder whether they would think again, or even have the confidence to stick their heads above the parapet. I would certainly welcome the company!

Even if words like 'spiritual' and 'sacred' are a step too far for some, anyone who stands back and considers what has been done to Nature by what is now the dominant approach could be forgiven for thinking that simple common sense has been abandoned. How else could we have embarked upon such a singular and self-destructive enterprise to prove beyond doubt that we can, indeed, do without the rest of the natural world? For that is what we are doing. We are testing the world to destruction and the tragedy – no, the stupidity – is that we will only discover the real truth when we have finally succeeded in completely denuding the world of its complex life-giving forces and eradicating traditional human wisdom.

If we continue to engineer the extinction of the last remaining indigenous, traditional societies, as is happening in so many countries today (where governments feel embarrassed because they make a country look less 'modern'), we eliminate one of the last remaining sources of that wisdom. For just as natural species, once lost, cannot be re-created in test tubes, so traditional, so-called 'perennial' wisdom, once lost, cannot be reinvented. This is the real damage being done by our disconnection, which is fast becoming all but complete in the modern world, all the while proving that the great experiment to stand apart from the rest of creation has failed.

This is why I have argued for so long that we need to escape the straitjacket of the Modernist world view, so that we can reconnect our collective outlook to those universal principles that underpin the health of the natural world and keep life's myriad diversity within the limits of Nature's capacity. In other words, we have to discover once again that in order for humanity to endure alongside the natural world (and the vast, as yet unnumbered creatures with which we share this miraculous planet) on which it so intimately depends for its survival, it is essential to give something *back* to Nature in return for what we so persistently and all the more arrogantly take from Her. Our approach cannot all be based on 'rights'. There have to be 'responsibilities' too. And my mentioning the word 'Modernist' brings me to one more term I need to define before we go any further.

Modernism

Every time I use this word it provokes a storm of protest. Perhaps it is because, for many, 'Modernism' conjures up a certain kind of popular 'trophy architecture' associated with, for instance, Le Corbusier, who famously described a house as 'a machine for living in'. However, the Modernism I am referring to is a much more pervasive doctrine than the eye-catching and clever style of architecture created by a complex figure like Le Corbusier. He was a Modernist, of course. He certainly subscribed to the wider principles of that movement – its devotion to the machine, its love of speed, the rejection of beauty as being innate in things, and the denigration of traditional design and craftsmanship.

What we should remember is that what became an international movement and a far-reaching attitude began as a gross indulgence by the one-time avant-garde. You have only to take a look at Marinetti's famous Futurist Manifesto, published in Paris in 1909, to see what I mean. Even he called it 'demented writing'. His language, though, rings with a certain familiarity – for instance, when he calls for 'the gates of life to be broken down to test the bolts and padlocks' or when he urges humanity and technology to triumph over Nature. Marinetti did, at least, admit that he wanted to 'feed the unknown, not from despair, but simply to enrich the unfathomable reservoirs of the Absurd'. But that ambition seems to have been conveniently forgotten as the Modernist ideology tightened its grip.

Marinetti's historic prospectus was one of the statements that induced the wave of Modernism that would sweep the industrialized world throughout the first half of the twentieth century. Later I will be more specific in my definition of Modernism because its impact on our general outlook has been so pervasive, but for now suffice it to say that this is the movement that still, to my mind, underpins what has become the Establishment view. Modernism deliberately abstracted Nature and glamorized convenience and this is why we have ended up seeing the natural world as some sort of gigantic production system seemingly capable of ever-increasing outputs for our benefit. Modernism compounded what had already become a general attitude in industrialized countries towards the natural world and, as that definition has become more predominant, so the view we have of our own role in Nature's process has been reduced. We have become semi-detached bystanders, empirically correct spectators, rather than what the ancients understood us to be, which is participants in creation. This ideology was far from benign or just a matter of fashion. The Marxism of the Bolshevik regime totally absorbed, adopted and extended the whole concept of Modernism to create the profoundly soulless, vicious, de-humanized ideology which eventually engineered the coldly calculated death of countless millions of its own citizens as well as entire living traditions, all for the simple reason that the end justified the means in the great 'historic struggle' to turn people against their true nature and into ideological, indoctrinated 'machines'. All this I will explain because the impact of the industrial mindset focussed by Modernism is key to the situation we face today. It is responsible for the loss of a deep experience of the interconnectedness of Nature, severing a meaningful relationship with the world we inhabit.

Making the shift so that we see things in a much more joined-up and deeply anchored way – the way things really are rather than as they appear to be – is the first stage of the Sustainability Revolution. But we must approach the challenge positively, regarding such a revolution as an opportunity rather than

Manufactured tower blocks rising above Dundee, Scotland, in the late 1960s. These buildings were built quickly to meet a housing need, assembled using a system of concrete panels made on a production line. They lasted just 40 years before they were deemed too old. This is not sustainable architecture in any sense, not least for the people unlucky enough to live there.

as a threat. We will all have to alter our outlook on life, but we could see this as an investment rather than as a tax. It will inevitably require a period of reassessment of our values and priorities and a realignment of approaches. But if it comes about, it must do so through interchange and discussion rather than by imposition or decree. It is my ambition that this book, the film that will follow it, and other initiatives that will accompany both, will help to facilitate that vital cross-cultural and international discussion and exchange.

The Yorkshire Dales where the buildings and walls are made from local materials, creating a unique landscape where the Man-made blends with the natural and works best for the harsh conditions found there.

My hope is that I have at least made it clear so far that in the twenty-first century we desperately need an alternative vision that can meet the challenges of the future. It will certainly be a future where food production and its distribution will have to all happen more locally to each other and be less dependent, certainly, on aircraft; where the car will become much more subordinated to the needs of the pedestrian; where our economy will have to operate on a far less generous supply of raw materials and natural resources. But it could also be one where the character of our built environments once more reflects the harmonious, universal principles of which we are an integral part. It could involve a way of teaching our children which offers a much more comprehensive view of reality – one which emphasizes our interconnected reliance on every other part of the whole and living system we call 'Earth'.

As it is, by continuing to deny ourselves this profound, ancient, intimate relationship with Nature, I fear we are compounding our subconscious sense of alienation and disintegration, which is mirrored in the fragmentation and disruption of harmony we are bringing about in the world around us. At the moment we are disrupting the teeming diversity of life and the 'ecosystems' that sustain it – the forests and prairies, the woodland, moorland and fens, the oceans, rivers and streams. And this all adds up to the degree of 'dis-ease' we are causing to the intricate balance that regulates the planet's climate, on which we so intimately depend.

My entire reason for writing this book is that I feel I would be failing in my duty to future generations and to the Earth itself if I did not attempt to point this out and indicate possible ways we can heal the world. I could not have contemplated producing it even two years ago, but I feel the time may now be more appropriate. I sense a growing unease and anxiety in people's souls – an unease that still remains largely unexpressed because of the understandable fear of being thought 'irrational', 'old-fashioned', 'anti-science', or 'anti-progress'.

We live in times of great consequence and therefore of great opportunity. This book offers inspiration for those who feel, deep down, that there is a more balanced way of looking at the world, and more harmonious ways of living. It will not only outline the kinds of approach that depend upon us seeing Nature as a whole, but also examine the great and practical value in seeing the nature of humanity as a whole. What I hope will become obvious is just how many answers we already have at our disposal, if our goal is to re-establish our rightful relationship with Nature and pull back from the brink of catastrophe. It is a goal I truly believe is achievable, if we remind ourselves of the essential grammar of harmony – a grammar of which humanity should always be the measure.

Nature   2

Attachment to matter gives rise to passion against Nature. Thus trouble arises in the whole body; this is why I tell you: be in harmony.

The Gospel of Mary Magdalene

I want to begin in Cambridge, where I was an undergraduate over forty years ago. One of the colleges at Cambridge University is called Peterhouse. It was founded in 1284 and is the university's oldest surviving college. Its medieval architecture is today among the most precious in a city that is unusually rich in striking historic buildings. Alongside portraits of past Masters of Peterhouse, the college hall is decorated with fabric hangings designed by William Morris. He was a leading inspiration in the nineteenth century for the romantic Arts and Crafts Movement, having set about rediscovering the lost techniques of embroidery, stained-glass window making, illumination and calligraphy, textile dyeing, printing and weaving. He did this as a defiant reaction to the roughshod commercial expansion of machine-based manufacturing of the time. Morris was not against machines. He was concerned about what machines were being put to do and was horrified by the human degradation of work in the nineteenth-century factory and how the land was being ruined by industrial pollution. He also lamented that art and beauty had no place in this factory-based world and felt that, as a consequence, human dignity lay in ruins. On the ancient stone walls of Peterhouse's hall his work seems very much at home.

Some time after Morris completed these designs and they were placed there, they were joined by more modern fittings, symbols of the age of industrialization that he was so concerned about. Most visitors today would hardly give them a second thought: they are the hall's electric lights. What makes them special is that they were the first to be switched on anywhere in the university. In the country as a whole they were second only to those in the Houses of Parliament the year before. They were installed at the insistence of the scientist John Kelvin to mark the six-hundredth anniversary of the college in 1884. The electricity for the lights came from a generator powered by a steam engine that was in turn powered by gas derived from coal.

The installation of electric lighting at Peterhouse marked a dramatic

PREVIOUS SPREAD:
The Imperial Valley would be a desert were three quarters of California's river water not allocated for agricultural purposes. 'Flood and furrow' is employed here, which experts say wastes vast quantities of precious water. To keep pace with demand, water is being diverted from the Salton Sea, which is consequently drying up.

LEFT: *1864 trellis wallpaper design, by William Morris (1834–1896). Morris was an artist associated with the English Arts and Crafts Movement and Pre-Raphaelite Brotherhood. Morris developed strong views on mechanization, the environment and on the need for conservation in both the natural and built worlds.*

Pioneer Run Creek, Titusville, Pennsylvania, USA, in 1865. One of the world's first oil fields. From small beginnings, nearly every aspect of our lives is now dependent on fossil oil and gas.

departure from the previous 600 years, during which the college hall had been lit by candles, oil lamps (most likely including those powered by the blubber of whales) and wood fires. The steady, yellowish electric glow that we now take so much for granted also signalled a new dawn, not only for the college but for the whole world: the age of electricity and mass-produced power had arrived. It marked perhaps the most significant turning point in human history, and what may indeed prove to be the most significant in the recent history of all life on Earth.

When those lights first illuminated the shady, cool recesses of Peterhouse's ancient hall, the world's human population had reached about 1.2 billion. Although horses were still an indispensable source of power, the steam railway that reached Cambridge in 1845 had already led to a revolution in attitudes to travelling. A year after the great switch-on at Peterhouse, the first patent for a petrol-engine car was granted to Karl Benz in Germany. Eighteen years later the first plane would take to the air. In the expansive fenlands surrounding Cambridge, much of the drainage system that had been dug in Roman times had for centuries been powered by the wind. But even in these remote East Anglian flatlands times had changed. Some sixty years before the switch was first thrown to light Peterhouse, the last of some 800 wind pumps had been closed down, to be superseded by steam-powered alternatives. The obsolete technology was quickly swept away in favour of more flexible and powerful modern alternatives.

I make these points in order to underline the recent rapid pace of change, and to remind us of how quickly we have reached the point where we are now. All of these wonderful innovations were developed with the best of intentions. Pioneers like Benz wanted to create wealth and well-being and to improve people's lives. But the huge benefits we have gained have not come without serious consequences. One of the biggest is on view up the road from Peterhouse, in a different part of Cambridge.

Inside the headquarters of the British Antarctic Survey is what is without doubt one of the most remarkable archives ever accumulated. Kept in a building with a temperature permanently maintained at below −20° Celsius are tube-shaped sections of ice. They are about two feet long and some four inches in diameter. Some are white, others translucent. Meticulously labelled and filed on racks, these frosty samples were taken from the core of a drill that had

Ice core data has enabled scientists to build up a detailed long-term picture of carbon dioxide, methane and oxygen isotope levels in the atmosphere, and how average temperatures have changed.

penetrated one of the thickest layers of ice found on our modern Earth: an area called Dome C, high on the plateau of East Antarctica. As snow accumulated over thousands of years, it built up in deeper and deeper layers. The snow was compressed by subsequent falls and the weight formed layers of ice, year after year for hundreds of millennia.

If the ice cores drilled out of Dome C were laid end to end they would stretch for about two miles, but it is not so much the depth the drill reached that is most remarkable: it is the age of the ice it found at the deepest levels. The further the drill went down, the older the ice it extracted, at the bottom pulling up samples more than 800,000 years old. The ice drill thus enables us to look back over the best part of a million years and so in a way is a time-travel machine.

Trapped in the falls of ancient snow are tiny bubbles of the Earth's atmosphere as it was when the snow fell. When I last visited the centre, one of the scientists took an ice sample into the warmer corridor, where it began to melt, and I could hear it crackle as the bubbles of trapped air were released. By catching and analyzing these little bits of frozen air, scientists at the British Antarctic Survey have been able to take direct measurements of different gases. By analyzing different oxygen isotopes in the ice itself they can calculate accurate temperature measurements and these confirm that during the past three-quarters of a million years or so our world has undergone eight separate ice ages, each punctuated by warmer interglacial periods, like the one we are living in today.

One of the most dramatic findings to emerge from the study of these ice cores is how the levels of carbon dioxide have changed, and how closely they have matched the climate changes. Even in the warmest periods, the ice cores show that the concentration of carbon dioxide never went above about 290 parts per million – that is 0.029 per cent by volume of the air. In the 1780s, when the Industrial Revolution began in England, that figure stood at what was then a recent high point of about 280 parts per million. But the ice cores show how this figure has increased, and fast. A hundred years later levels had crept up to just below 300 parts per million. Another hundred years later, in the 1980s, carbon dioxide levels exceeded 340 parts per million.

Today there are more than 50,000 large power stations in the world, most

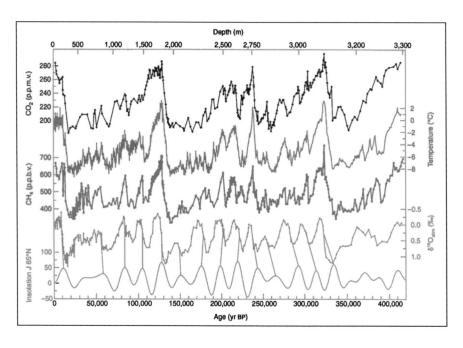

Ice core data has enabled scientists to build up a detailed picture of carbon dioxide, methanol and oxygen isotope levels in the atmosphere and how average temperatures have changed.

HARMONY

of them burning fossil fuels to do what was done at Peterhouse in the 1880s –
but now on a vast global scale. Between then and now the emissions we produce
by generating energy in this way have grown from virtually nothing to billions
of tonnes of carbon dioxide every year. And as the demand for power increases,
so the number of coal- and gas-fired stations also goes up. In China and India
today the process of industrialization is happening faster than ever before. China
is opening one or two major coal-fired power stations per *week*. New cars are
rolling onto China's roads at a rate of about 1,000 a day. Of course, China and
other countries are simply following a pattern of development seen in the West,
but what I have been told by the climatologists makes it very clear that our
collective way of life simply cannot continue like this if we are to avoid
catastrophic impacts on our planet's climatic stability.

Ice samples are a vital source of information, but for the past fifty years or
so our knowledge of changing carbon dioxide levels has also come from direct
measurements of the composition of the air. High on Mount Mauna Loa in
Hawaii, as far away as possible from carbon dioxide sources such as forest
clearance and industry, scientists have been able to monitor global change taking
place before their very eyes. The measurements taken in Hawaii do not reveal
the kind of slow change that occurs over tens of thousands of years, but a more
rapid trend which is visible year on year. In 2010 the samples taken in Hawaii
revealed that carbon dioxide levels had climbed above 388 parts per million.

What is perhaps even more alarming than the very high level of carbon

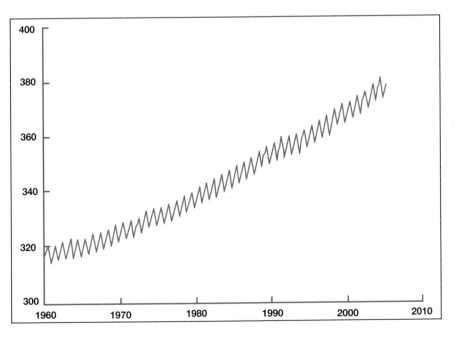

Samples taken at the observatory on Mount Mauna Loa, Hawaii, since the late 1950s have revealed the fast-changing nature of Earth's atmosphere. This famous diagram is known as the Keeling Curve after scientist Charles David Keeling who initiated this programme of measurement.

dioxide is the rate at which it is increasing – for not only is it going up, it is going up at a faster pace. This is not solely because our consumption of fossil fuels continues to grow and that we continue to destroy vast areas of forest every year, but also because the natural systems that have so far absorbed most of the carbon dioxide we have released are now apparently doing so more slowly than they did in the recent past. It could also be because some natural systems, such as soils, are beginning to release carbon dioxide as a consequence of the warming that has already occurred.

Unfortunately, the emissions we create do not just conveniently disappear somewhere into space through strategically placed holes in the sky. They are trapped within the Earth's atmosphere. This has always happened and there are various systems that can help correct any imbalance. When it comes to the climate one such system is the oceans. But such is the sheer scale of our emissions that this particular safety valve is becoming overwhelmed. Also, worryingly, the vast quantities of carbon dioxide absorbed by the oceans are changing the water's composition, making it more acidic.

The effect so far is quite mild, but it will increase as the level of carbonic acid in seawater increases. It is a simple chemical relationship. More carbon dioxide in the atmosphere leads to more acidic seas. And if the sea is more acidic, then all sorts of creatures will find it harder to survive. Greater acidity, for instance, affects the ability of shell-forming animals to grow, which in turn could unbalance whole ecosystems, like coral reefs, and disrupt the food chains that support bigger animals, like fish and birds and, of course, us, because we catch the shellfish and fish and eat them. Also, if shell-forming creatures cannot function properly because of changes in the ocean's chemistry, this could also have implications for the state of the atmosphere. Once they stop making shells, they remove less carbon.

This is just one of many thousands of examples that demonstrate how everything in Nature is connected – it works as a whole system, with changes in one part affecting the behaviour of others. As we change the atmosphere, so this causes changes in the oceans, bringing about ecological changes that in turn will potentially compound the speed and scale of changes in the atmosphere. The recent science seems to me to confirm what many people have felt for some time: namely that the relationship we have developed with the Earth during our technological and industrial age is a one-sided affair. It is not one of equals, but is based upon the master-and-slave model, and many see this as posing grave threats not only to the natural world, but to humanity as a whole.

The inconvenient fact is that the changes we have relatively recently set in train on Earth are occurring faster than any in human history, and on a larger scale. These changes have enormous implications for people everywhere. I believe it is necessary for us to think very seriously about how we curb our impact before it is too late. We might begin by looking in particular at how we use energy and land.

When it comes to energy, our world today depends on fossil fuels. From cars to fridges and from computers to airliners, every aspect of how we live now depends upon fossil fuels. Fossil energy has not only allowed the development of thousands of vast cities, but has also permitted the industrialization of farming and created the wealth to invest in tremendous advances in healthcare. Both these have enabled our population to climb towards seven billion and it continues to rise rapidly.

Fossil sun

Giant ferns that grew in carboniferous swamp forests 300 million years ago and microscopic plankton that once drifted in the oceans, were among the organisms that helped to create the coal, oil and gas deposits that now drive the modern world. Provided by the sun, energy that flowed through living organisms long ago in the Earth's past was trapped, transformed and then held tight in the hydrocarbon compounds that now power our industrial societies. That energy source, which took hundreds of millions of years to accumulate gradually, is being used up all at once, and the carbon that was, in the process, taken out of the air and stored away over such vast periods is being released as carbon dioxide in a great single pulse. As a result, we are upsetting a balance established over millions of years, and on a massive scale. This is an indisputable fact and it seems to me that the ice cores and the graphs drawn from the Mauna Loa measurements act as a cold, silent and independent witness to the changes that have occurred after more than two centuries of industrialization.

However, it is not only fossil fuels that are causing an ever more pressing atmospheric crisis. Around a fifth of annual human-induced carbon dioxide emissions come from the continuing clearance of our planet's forests, particularly the tropical rainforests. During 2007 I received the direst warnings from leading scientists as to the implications of this particular trend.

The tropical rainforests are to me without doubt the most incredible terrestrial ecosystems on Earth, and for a considerable number of years I have been working hard to convey this fact to as many people as possible. They are literally the Earth's air-conditioning system, and are also a vital mechanism that moves water around the globe. Without them we would have less of the rain that, for now, allows crops to grow and farms to thrive, but they are also enormous carbon storage reservoirs. Billions of tonnes of carbon are locked up in them, but as the rainforests are degraded and cleared away by logging operations, or replaced by cattle pasture, or with plantations of commodity crops, such as soya and palm oil, the carbon that was once in the trees and the soil beneath them is transferred to the atmosphere as carbon dioxide, adding to the burden of greenhouse gases.

What makes matters considerably worse is how the tropical rainforests also absorb about 15 per cent of our emissions, such as those from cars and power stations. As they are cleared, so their ability to reduce the impact of our emissions from industry and transport is diminished. More than a third of the tropical rainforests have already been removed, the vast majority since the 1950s. The rest are fast disappearing, as demand for land to grow the crops that feed global commodity markets spirals upward. This is for me one of the greatest tragedies of our age, not least because it is unnecessary.

Necessary or not, the increased levels of carbon dioxide brought about principally because of fossil fuel combustion and deforestation have already led to an elevated global average temperature. Scientists are now confident that the net effect of human activities since the mid-eighteenth century, when coal use began to rise rapidly, has been one of warming. Global average temperatures

have increased by about 0.8° Celsius since this time and there is strong evidence that most of the observed increase in temperature since the mid-twentieth century is directly attributable to increases in anthropogenic carbon dioxide and other greenhouse gases. This is now a mainstream view, backed by a vast body of scientific evidence.

Of course the changes we have caused to our planet's atmosphere were not an intentional or conscious act. They are the inadvertent results of how we have grown economies, made life more comfortable and promoted people's welfare. They are unintended consequences of what we have come to call development, and the process continues today as we seek to expand our comfort and solve problems in ways that generate yet more greenhouse gases. But it is different now, not least because I don't believe we can go on any longer pretending that we do not know about the problems we are creating. Thanks to modern science we now know the consequences of the way we have chosen to live, and the way we have become used to meeting our needs – and we are also increasingly aware and informed about the grave legacy we are leaving for those who follow us in the near future.

OVERLEAF: *Moon rise over an oil refinery. Refineries heat crude oil and pass it through catalytic cracking towers. Pollution released in the process causes the moon to appear red.*

Melt

What I have been told by scientists for years has finally been accepted as common knowledge – at least among those prepared to face the facts. From humanitarian disasters on a grand scale to extreme conditions causing massive costs to the insurance industry, and from the inundation of low-lying coastal areas to the disappearance of rainforests owing to prolonged droughts, we could soon witness profound challenges that could become unmanageable if we do not act in time. The consequences for hundreds of millions of people living in developing countries will be disproportionately worse – although, for many, they already are. Having said this, it is not only the poorer nations that will suffer the short-term shocks caused by extremes in the weather. The devastation wrought on New Orleans by Hurricane Katrina is a recent case in point. I saw with my own eyes the impact of that catastrophe – and it is sobering to remember that it occurred in the world's richest economy.

My wife and I visited New Orleans about two months after Hurricane Katrina hit the city. We were appalled to see the extent of the damage caused there. This event demonstrates how we are all at risk, rich and poor alike.

In addition to more extreme conditions, including severe floods, droughts and heatwaves, there is good reason to be seriously concerned about what in the jargon is known as 'non-linear' change. This basically means change that is not smooth and gradual, but sudden and dramatic. One example that is taking place right now before our very eyes is the rapid loss of the Arctic ice sheet. I recently heard from the polar explorer, Pen Hadow, how fast this is happening – much faster than scientists predicted even a few years ago.

In the late Summer of 2007 an area of the Arctic sea ice twice the size of Great Britain disappeared over a couple of weeks – an event never before witnessed. The same thing happened in 2008. On the basis of recent events, and rates of warming in the Arctic region which are far higher than the global average, some believe that in twenty to thirty years the Arctic Ocean could be virtually free of sea ice in Summer. Some projections suggest that this will happen even sooner. There are also signs of rapid change in Antarctica. The British Antarctic Survey recently explained to me that, out of 244 glaciers on the Antarctic Peninsula that have been monitored for the past fifty years, some 87 per cent of them have been in retreat. The Greenland ice cap is also showing signs of increasingly rapid melting – and should the whole lot melt it will eventually add something like seven metres to the average global sea level.

The loss of ice is also serious because ice acts like a mirror reflecting the Sun's energy back into space. When it is gone, it is replaced by darker ocean and land beneath, which absorb some 90 per cent of the energy coming in, whereas the ice would have reflected 80 per cent of this back again. This effect is known as a positive feedback, whereby the warming leads to changes that cause more warming, no matter what we do. And unfortunately this is not the only effect. The warmer our atmosphere gets, the more numerous and dangerous the different 'feedbacks' are expected to become. These range from the melting of permafrost, which leads to the release of billions of tonnes of carbon dioxide

This image shows the extent of sea ice in the Arctic at the end of the melt season in September 2008. In recent years there have been record levels of open ocean as the ice has melted to new low points. This rapid large-scale change threatens both wildlife and the traditional livelihoods of indigenous people.

and methane (a very potent greenhouse gas) to the billions of tonnes of greenhouse gases that would also be released as rainforests die back. These are, to leading climate scientists around the world, signs of global imbalance.

As far as climate change is concerned, there is now a strong consensus that, in order to avoid the worst consequences of what we have put in train, it will be necessary for us to limit global average temperature increase to no more than two degrees Celsius above the average temperature prevailing at the end of the pre-industrial period – that is, in the late eighteenth century. A two-degree change between then and the near future may not seem very much, but as a shift in the global average it is a vast change, and in some areas it will be a lot bigger – in the Arctic, for example.

Temperatures even three degrees above the pre-industrial average have not been seen since the Pliocene period, some three million years ago. At that time sea levels were up to 25 metres higher than now. The last time our planet saw a comparable four degrees of warming was millions of years before then, and five degrees perhaps some 55 million years ago, at the beginning of the Eocene. At that time the Earth was virtually ice-free, with the average sea level some 75 metres higher than now. Animals such as marine turtles that today we associate with the tropics and sub-tropics lived at polar latitudes. A world that is on average four degrees warmer than now would be very different, and would undoubtedly bring very major upheavals and create major risks for most of humanity.

I recently visited the Met Office Hadley Centre for climate research at Exeter, in Devon, England, where scientists confirmed to me that a four-degree temperature increase during the second half of this century is to be expected on the basis of 'business as usual' emissions, should we not act now. However, it could be higher than this with the worst-case scenario for the end of the twenty-first century, suggesting it is conceivable by then that average global temperatures will rise by over six degrees. To put that into perspective, a rapid six-degree temperature increase is what some scientists believe to have brought about the mass extinction that marked the end of the Permian period some 250 million years ago. That six-degree increase is estimated to have taken some 10,000 years to occur and it took life on Earth tens of millions of years to recover from that catastrophic episode of global warming. Back then the climate change was probably caused by volcanic activity: this time the warming is caused by power stations, deforestation, cars, farms, and factories. If we don't do something about these sources of global-warming gases very quickly, then we could, the most recent projections show, trigger six degrees of warming not over 10,000 years, but in less than a century.

Temperatures even three degrees above the pre-industrial average have not been seen since the Pliocene period, some three million years ago. At that time sea levels were up to 25 metres higher than now.

At the moment this appears to be where we are heading. Emissions scenarios prepared in 2000 set out different possible future emissions levels based on different assumptions about economic growth and the uptake of new technology. And yet in 2005, 2006 and 2007 the world's emissions of greenhouse gases were *above* the line set out in the *worst-case scenario*. We have arrived at the brink of potential disaster, and yet we still accelerate towards the edge.

Glaciers are very sensitive to climate change and as the average global temperature rises glacier retreat is underway worldwide. At Glacier Bay National Park, Alaska, there has been dramatic glacier shrinkage. On the left a steamship sails toward the Muir Glacier in 1902. From the same viewpoint in 2005 the glacier has disappeared.

Nevertheless there is some cause to be encouraged. We still have some time to minimize the impact of the increases in carbon-dioxide levels, but not much. For the industrialized developed countries, a cut in carbon-dioxide pollution of about 40 per cent by 2020 might place the world on an emissions reductions path that could avoid the worst effects of what lies ahead. That is not to say that the impact we have on the world could or must be reduced to zero by then, but rather that this is the direction we should be travelling in. What is certain is that this is mostly the opposite direction to that travelled by humanity today.

The evidence which shows Man-made impact on the climate system recently became the subject of fierce debate. Those who are sceptical that humanity's activities have caused climate change made an all-out assault on the evidence base in the run-up to the UN Conference on Climate Change in Copenhagen at which I spoke. The row had an alarming effect on public opinion, and being so close to it I could see that it was a deliberate attempt to dampen the justified concerns about the climate change threat that the Media had been increasingly reporting in recent years. But whether you choose to believe the sceptics rather than the vast body of evidence that now exists, climate change is not the only large-scale ecological challenge that we face. Alongside changes to our planet's atmosphere, the activities of our single species have caused an unprecedented assault on the fabric of life.

Depletion

Modern scientific investigation tells us that today we sit at the pinnacle of some 3.5 billion years of evolutionary refinement. It seems to me that the interdependent web of connections, relationships and flows of energy, the finely woven tapestry of life, is undoubtedly the greatest marvel ever placed before us.

It is not necessary to travel to the tropical rainforests or to dive among tropical coral reefs to be at the cutting edge of our knowledge of life on Earth. In 2008 a scientist working at the Natural History Museum in London was presented with what turned out to be possibly a 'new' species of bug. It was picked up by his son while eating his lunch in the museum's central London gardens. The almond-shaped insect, about the size of a grain of rice, had made itself at home in the sycamore trees in the nineteenth-century museum's grounds. About a mile away at Buckingham Palace, two species of mushroom, apparently new to science, were recently found in the gardens.

I have for many years been privileged to spend time speaking with and learning from some of the world's leading experts on the natural world. Today they have a new name for this tremendous variety of life: it's called biodiversity. The most incredible fact about the multiplicity of forms of life and the myriad associations they all form is the amazing variety. The most alarming is the rate at which it is disappearing. There are many stories that underline the urgency we face in stemming the tide of biodiversity loss that is taking place all around us. One that is close to my heart begins high on a hill in New Zealand.

Taiaroa Head is at the northerly tip of the Otago Peninsula, overlooking the entrance to Otago Harbour, at the head of which is the town of Dunedin. Here in Spring it is possible to watch at close quarters one of the most remarkable of all birds: the albatross – more specifically the Northern Royal albatross. The birds that nest here each year make up the only mainland albatross colony in the world. All the others are on small islands, with many located on some of the world's most inaccessible oceanic outposts.

The albatrosses that breed at Taiaroa are enormous creatures. White, with great long black wings, they are surely the most majestic of birds. They nest in peace because they have the full protection of New Zealand law. They are a major tourist attraction and have made Dunedin world-famous. But when they have reared their single chick and set out to sea once more, they face great peril. For millions of years albatrosses have wandered the oceans as masters of wave and wind. Twenty-four kinds of these great birds hunted and roamed in a watery world apparently without limit. During my time sailing the high seas when I served with the Royal Navy nearly forty years ago, I watched these birds, marvelling at the vastness of their world and the way they were perfectly adapted to the conditions that enabled them to thrive in what for us humans are such hostile conditions. But even this great ocean wilderness is no longer the

The Taiaroa Head Royal Albatross colony at Dunedin in New Zealand is an unforgettable place to visit. This six week old chick is doing well.

This young Laysan Albatross died after being repeatedly fed with plastic debris collected from the sea by its parents.

sanctuary it was for these mariners of the remotest seas. No fewer than twenty-one species of the twenty-four are now regarded as being in danger of extinction.

This is because it is not only albatrosses that seek a living from the vast and seemingly empty oceans that ring Antarctica. Thousands of miles from their home ports, long-line fishing boats also patrol these wild seas. They come after large predatory fish and catch them with hooks trailed out on lines up to an astonishing eighty miles long carrying tens of thousands of small squid for bait – exactly the food of albatrosses. The hooked lines aim to catch valuable species such as toothfish, tuna and swordfish, but the baits are also lethal for albatrosses.

The birds dive onto the baited hooks and the sharp barbs slice into their beak or throat. A vain struggle is soon followed by a painful death by blood loss or drowning. It is an unglamorous end for birds that, during a lifetime that can span sixty years, will have travelled considerably further than the distance to the moon and back. The plight of the albatrosses has recently emerged as a conservation cause célèbre, and I have been pleased to help BirdLife International and others to make the case for a change of fishing methods, but despite some efforts to protect these incredible birds, long-line fishing boats continue to take a terrible toll. It is a situation, though, that as in so many other cases is complicated by the ability of countries to cooperate internationally, and is not helped by the fact that the USA has still not signed the International Agreement on the Protection of Albatrosses and Petrels. And alongside the decline of the albatrosses has come the near destruction of the populations of the large fish most prized in international markets. There is an awful congruity

A long-lived Patagonia Toothfish is gaffed aboard a long-line fishing boat.

in the fact that the Chilean Sea Bass that these boats catch also live for sixty years and, like the albatrosses, they do not begin to breed until they are between 10 and 20 years old. If they are caught before that time the chances of any future stock are gone for good.

While the fish have become scarcer, the rewards to be gained from catching them have only increased. Controls are in place to help avoid the complete collapse of the populations of the most valuable species, but illegal fishing has become a major problem. Even with the best will, policing the vast expanse of ocean that is plundered by illegal fishing presents a massive challenge.

Given the low numbers to which many albatross species have already sunk (and some hang on as just a handful of remaining birds), it is obvious where these trends will take the world's most charismatic and endangered sea birds. This situation is made worse still for some species because not only are they attracted to baited hooks, but they also go for pieces of discarded plastic. The adult birds pluck plastic items from the sea, mistaking our consumer detritus for food, and then they feed it to their young – which often kills them. Recently I was told of the vast rafts of plastic that now float in the Pacific Ocean some 500 miles off the coast of California. This appalling phenomenon of today's world has doubled in size over the past decade and it now occupies an area of 540,000 square miles of the Pacific – nearly six times the size of the United Kingdom. This 'plastic vortex', as it has become known, comprises up to 100 million tonnes of man-made waste – plastic packages, bottles, cans, tyres and broken-down chemical sludge. This monstrous plastic island is situated in a relatively stationary region of the ocean because it is bounded by a system of rotating oceanic currents called the North Pacific Gyre. Its proportions are truly eye-watering and it puts the plight of these amazing sea birds, as well as all kinds of marine life including turtles, into its proper perspective, not least because hardly anybody seems to know about the problem so nothing is done about it. Call me a 'busybody' but I am determined to do all I can to make sure this is not the case for much longer…

On land there are just as many problems. Perhaps surprisingly, one of the main reasons for the extinction of many creatures over the past few centuries has been the sometimes subtle impacts caused by introduced species. Rats, for example, have found their way to remote oceanic islands courtesy of explorers' ships. When aggressive generalists like these turn up in ecosystems where the native animals have evolved in isolation and do not have the defences to deal with alien invaders, mayhem can result. Indeed, most of the recorded extinctions that have occurred since 1600 have been due to the effects of introduced exotic species. The dodo, the very emblem of extinction, appears

The native Red Squirrel has disappeared from most of its former range in Britain.

to have succumbed more to the pigs and monkeys released by sailors than it did to excessive hunting.

In the UK we welcome North American visitors with open arms – except perhaps for one: the grey squirrel. This charming-looking crea-ture has a dark side. It has devastated our native population of red squirrels, hastened the disappearance of our native dormouse, and in some places has caused decline in songbird popu-lations. It also causes untold damage to young hardwood trees such as oak, beech and ash. These little creatures are frustrating so many worthy efforts to re-establish native hardwood plan-tations, and they provide one more example of how our interference with natural systems can cause chaos.

I don't suppose we will be able to do much about the grey squirrels, save controlling them locally when their numbers grow, but my heart sinks

every time I hear of the latest big idea to introduce yet another new species of animal or insect from elsewhere in the world in order to deal with a problem caused by a previously imported species. And I am completely exasperated when such schemes from time to time are given the blessing of wildlife groups. Clearly even some of those who work closely with Nature, and who struggle so hard to protect her, sometimes think with the same mechanistic ideas that created the very problems they are trying to solve.

There is some good news, however. The Red Squirrel Survival Trust is for example, working hard to maintain and expand the populations of this wonderful creature. I have been pleased to help them with their effort to save these utterly charming creatures from what looks like imminent oblivion in these islands.

Elimination

While it is sometimes hard to comprehend the scale of the impact we have had on the Earth, the picture pieced together by the meticulous work of thousands of scientists tells an increasingly worrying story. Many in conservation circles believe that a sixth great extinction event is under way – a situation that might soon lead to a tsunami of species loss. This is not least because the rate at which species are being lost is now estimated to be between 100 and 1,000 times the natural background rate at which species disappear. Different kinds of animals and plants have always disappeared, but at the current high rates some projections suggest that by the end of this century we could lose up to 50 per cent of the total number of species that now inhabit the Earth.

Of course, during the long period that life has existed in abundance on Earth there have been times of rapid change. Indeed, etched into the fossil record are five periods when a large-scale loss of animal and plant species occurred – the last was when the dinosaurs disappeared, about 65 million years ago, marking a sharp boundary in the geological record between one epoch and another. Many believe the world recently entered a new one. And this one marks the fact that for the first time in the history of the Earth a single species has become the most dominant ecological agent – it is we humans, and that is why the period is called the Anthropocene. We are now the main reason for the rapid erosion of natural diversity, and whether we like it or not, this great living powerhouse is what sustains our well-being. We deplete and degrade it at our peril. In the pages that follow I will set out some of the reasons why this is the case, but for now it is enough to say that one reason why we are losing natural diversity so quickly is the rapid increase in our numbers.

In 1900 the world population was about 1.6 billion. By the time I was born in 1948 it stood at 2.6 billion people. By the end of the twentieth century it had reached over six billion – marking a near fourfold increase in 100 years. In 2010 it will have exceeded 6.8 billion and is expected to continue climbing inexorably to about nine billion by 2050. While our numbers rise, and as we become richer and thus demand and expect more, the ability of our planet to meet our needs has significantly decreased. We are using up its natural capital and resources as if they were inexhaustible and without long-term value.

So far, as this has happened, food supply has managed to expand to keep up with demand. In part this has been achieved by ever more dependence upon intensive methods of farming, which we will explore later, and in part by the conversion of more and more land taken from Nature and put to agricultural use. Keeping pace with this demand has come at a terrible cost and unless we adopt some quite fundamental changes in expectation and practice, the cost is set to increase. Taking the world as a whole, in 1900 there were 7.91 hectares of land per person, whereas in 2002, owing to the increase in population and loss of land to urbanization, that figure had shrunk to 2.02 hectares. In 2050 it is expected that there will be about 1.63 hectares each – and from that ever smaller plot of land we will demand more and more.

Much of the world's farmland is already virtually devoid of once native species of animals and plants, as modern high-tech agriculture has now basically

Amphibian species are being lost at a terrifying rate. Fortunately the strawberry poison-dart frog remains common throughout its Central American range.

turned farming into an arms race against Nature, excluding everything from the land except the highly bred crops designed to be resistant to powerful pesticides and grown using industrial production methods. These farming practices have a profound impact on the health – indeed existence – of the natural wildlife in the fields: the animals, plants, birds, the insects, microbes and bacteria that provide the services that sustain life. You don't have to conduct major scientific studies to see that this is so. Watch the plough as it is pulled through the soil in the fields of many modern industrialized farms and notice how few birds follow the tractor. The great clouds of birds that were once the normal entourage of the plough were only ever there for the worms: and with fewer worms and other invertebrates in the soil, there are fewer birds. Some soils have lost their ability to restore fertility naturally and have become little more than a media for growth through the application of chemicals. The process not only continues but intensifies, with the use of ever more sophisticated chemicals and now genetic manipulation too.

The prairies of North America are a dramatic example. The land there was transformed in a very short time indeed and on a continental scale. Some 75 per cent of the United States' original mixed-grass prairie has disappeared, much of it ploughed for the production of vast fields of maize, soya beans and wheat. This 'thin-skinned' land is now rendered at least temporarily productive with chemicals and machines, but in the opinion of many it should not be ploughed on such a scale. Its natural conditions render it vulnerable to collapse, which is what happened over large areas in the 1930s, when dustbowl conditions became one of the most potent images of the Great Depression. This situation arose through a combination of drought and ploughing; conditions that many believe could soon very easily be repeated – or exceeded – because of the effects of global warming.

It was the excessive exploitation of the land, natural resources and animals of North America that caused Aldo Leopold to observe in his 1953 book *Round River*: 'The outstanding scientific discovery of the twentieth century is not television or radio, but rather the complexity of the land organism. Only those who know the most about it can appreciate how little we know about it. The last word in ignorance is the man who says of an animal or plant: "What good is it?" If the land mechanism as a whole is good, then every part is good, whether we understand it or not ... who but a fool would discard seemingly useless parts? To keep every cog and wheel is the first precaution of intelligent tinkering.'

Clearly European colonists did not understand the 'land mechanism' when it came to the prairies – or if they did they chose to ignore what they knew.

LEFT: *Soybean harvest, Brazil. Ever more intensive methods have increased yields. In 2010 Brazil and Argentina, the two biggest exporters after the US, increased soybean production by about a third. Industrial farming methods cause major environmental impacts, however, ranging from increased greenhouse gas emissions to large-scale deforestation and from water pollution to biodiversity loss.*

Equipped with a philosophy of pioneering exploitation, they progressively plundered the land. In response to our attempts to keep pace with ever more demand for food, so the trend continues and, as a result, high up on Mauna Loa in Hawaii, scientists measuring changes to carbon dioxide levels are also finding other signs of global change.

Each year as China commences ploughing, so dust particles appear a while later at the mountain-top observatory – even though the fields and the monitoring equipment that is detecting the dust are thousands of miles apart. It is a sign of trouble. The dust indicates the loss of soil on a grand scale as a process of degradation and desertification unfolds in a manner that threatens real challenges for future food security. Top soil that is in many places just a few inches deep is perhaps our most precious resource, and yet, because of how we farm, vast areas of soil are being degraded or lost. Dust storms are more and more common in China, with some dust travelling to Korea and Japan as well as over the Pacific. According to the United Nations some 400 million Chinese live in areas threatened with desertification. Farming, grazing, deforestation, and irrigation methods have all helped to create giant dunes up to 400 metres high. Each year these shift inexorably forward, swallowing everything in their path.

It is not only soil that is increasingly on the move: so is our food. As a result of ever more globalized trade, food travels further and further, often covering vast distances between continents, even food that could be produced locally. In fact it has got to the point now where some countries effectively import huge areas of land, in the sense that farms in distant countries produce only food that will be consumed elsewhere. This adds to the threat posed to some countries' food security as land is moved from small-scale production for local markets to large-scale monoculture production to meet international demand. In the short term, countries might hope to earn revenue from these exports, but the costs of that policy can include serious social and economic problems, for example as water and land are driven into short supply and as food production for local consumption is reduced.

In the UK, our food in total travels an amazing eighteen billion miles each year. This includes imports by ships, trucks and planes. This produces an estimated nineteen million tonnes of carbon dioxide every year. Over two million tonnes of it is produced simply by cars travelling to and from shops. The majority of produce in the US travels between 1,300 – 2,000 miles from farm to consumer.

Perhaps the central message we should draw from this juxtaposition of circumstances is the need to promote sustainable farming everywhere, not just for local reasons, but for global ones too. Intensive monocultures destined for global commodity markets and dependent upon vast amounts of petrochemicals is a style of farming which, according to a recent UN survey, has been a major factor in causing up to a third of the world's farmable soil to be classified, to different degrees, as degraded – and that is in the past half-century alone. This is not a sustainable form of agriculture, and just so that we are clear, the dictionary definition of that word 'sustainable' is 'to endure without failure'.

In the UK, our food in total travels an amazing eighteen billion miles each year. The majority of produce in the US travels between 1,300 – 2,000 miles from farm to customer.

A truly durable farming system – one that has kept things going for 10,000 years – is the one that is commonly called 'organic farming'. In a sense this is an unfortunate term because it has the ring of an alternative approach, or even a new one, when it is actually how farming was always conducted before industrial techniques came to dominate agriculture. It means farming in a way that preserves the long-term health of the soil, which comes down to giving back to Nature organic matter to replace what has been taken out. It means maintaining microbes and invertebrates in the soil and good moisture. It means using good water catchment management, planting trees that prevent the soil being eroded and maintaining the teeming biodiversity, including the beneficial and essential insects, such as bees.

It has become ever more apparent to me that the 'food miles', the degradation of soils, the chemical pollution and the massive consumption of oil and natural gas add up to a way of producing food that acts without any concern for the harmony found in Nature and the natural order. This approach abandons the fundamentals that should sustain food production. But it is the increasing demand for land that poses the biggest challenge.

I should be clear, though, that this is not in my view merely a competition between technological approaches and different methods of farming. Once again it comes down to that fundamental question we will address in the following chapters of how we have been persuaded to look at the world and regard our place within the great scheme of Nature. Looking at the way we treat the natural world and produce our food raises questions far deeper than those of how it will be possible to save charismatic birds like albatrosses or to grow food without destroying the land.

Forest

We are all personally involved. Take what is happening in the middle of remote rainforests. Thousands of species are being destroyed every year as large areas of natural habitat are cleared to make way for farmland. This may seem so far away that it does not touch us. And yet a quick glance along an average supermarket shelf reveals that this is not so at all. From the supply of coffee and beef to the soya and palm oil that are ingredients in a huge range of processed foods, our modern world is presently fed in a highly destructive manner.

New roads are being driven into ever remoter areas of the rainforests to keep pace not only with the demand for food, but also to extract minerals and timber. A case in point is a new road that links Boa Vista in Northern Brazil with the Atlantic coast of South America at Georgetown in Guyana. It cuts through some of the most diverse and undisturbed rainforest in the world, and in doing so it threatens to unleash massive deforestation, which will in turn produce carbon-dioxide emissions and cause both the loss of vital biodiversity and huge disruption to the culture of the indigenous societies who still live there. The President of Guyana has told me how he sees little option but to allow the opening of the remote interior forests. His country needs income and the international community presently places a clear financial value on soya, beef, and timber, and so his country must clear the space to produce it. I have been very concerned about this but I am pleased to say that, in part, as a result of work undertaken by my Rainforests Project in bringing people together, a ground-breaking deal between Guyana and Norway has been concluded. The new agreement sets out to cut forest loss through providing Guyana with alternative economic strategies that will promote and enable low-carbon development, rather than the destructive changes in land use that have occurred in some other countries in that region.

It is not only in the tropics that the last wild forests are under threat. Even in the European Union some of the most extensive natural forests that remain are being cleared, especially in the East. In Romania, for example, where I have travelled extensively, agricultural expansion is leading to the clearance of wild forest. The forests are also being plundered for timber in ways that are quite unsustainable, in part because of the way in which the forests have been privatized, so that they lack proper governance and management.

Giant sequoias at Kings Canyon National Park. The world's largest tree species is confined to the western Sierra Nevada in California, USA. The tallest specimen alive today tops more than 270 feet. The oldest giant sequoias are in excess of 3,500 years old. They were well established at the time Tutankhamun was Pharaoh in Egypt.

OVERLEAF: *Farmers work on the rice terraces in Guiyang, Guizhou Province, China. This traditional method of farming on slopes helps to conserve both water and soil, making it a more sustainable form of food production.*

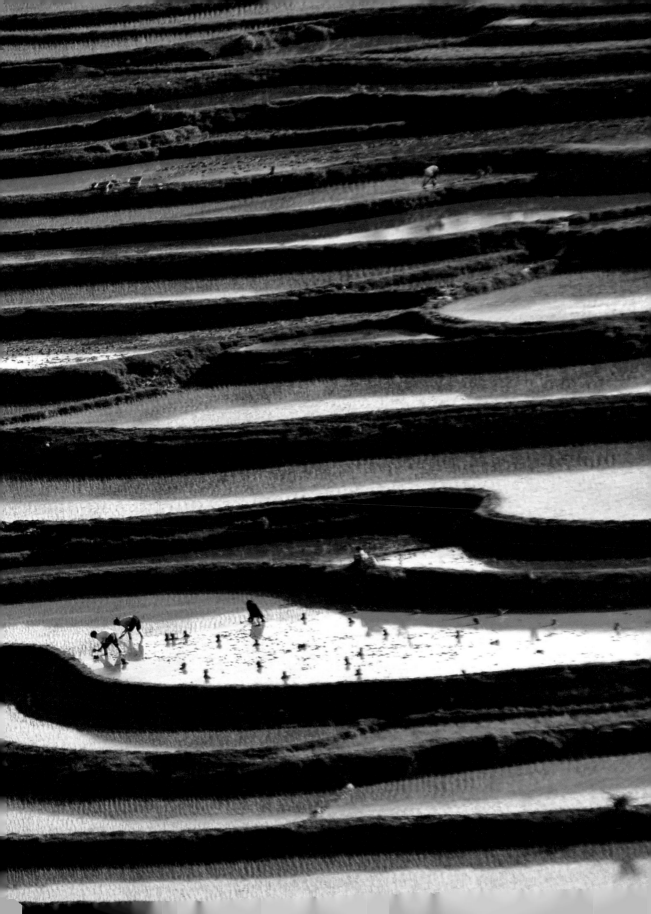

These trends have alarming implications. Human activity has already altered by different degrees nearly one half of the Earth's land surface. In the next thirty years this proportion is expected to rise to above 70 per cent. I learned recently that by 2050 it is expected that a further 11 per cent of land worldwide will be converted from natural habitats, either to become farmland or to become urbanized. That is about equivalent to the area of Australia.

Thou, Nature, art my goddess; to thy law
My Services are bound.

WILLIAM SHAKESPEARE

And to compound the situation, the relative intensity with which it will be farmed is also expected to increase dramatically. This, in turn, will cause a range of related problems. The rampant use of artificial fertilizers will continue to cause an excessive build-up of nutrients. This is already causing huge problems for the way natural ecosystems function, from grasslands to marine environments. These and other systems are also struggling to cope with the large quantities of manure produced in intensive livestock-rearing units. It may seem contradictory with so much talk of reduced fertility, but the progressive enrichment of the environment by animal waste, artificial fertilizers, and industrial sources of nutrients is one of the most serious ecological challenges we face. More and more fertility does not produce balance.

I have no doubt that taking land that at the moment is under the rainforests and converting it into ever more intensively productive farmland might meet the short-term demand for food and so create a source of income and development, but we should pause to remember that the biodiversity that we are depleting with such abandon is of fundamental importance to our welfare. Many of our modern medicines, a great number of industrial and chemical applications, and all of our food are derived from species that were once wild or still are.

As someone who takes an immense interest in farming, I am acutely aware of how people have shaped Nature to meet human needs. Working with the raw material of natural genetic diversity, farmers have, for thousands of years, conducted selective breeding, in the process creating and maintaining an incredible array of agricultural biodiversity. Some 10,000 years ago Mexican farmers began to domesticate maize from a wild grass. Since then they have generated thousands of varieties suitable for cultivation in the numerous environments in the Mexican landscape – from dry, temperate highlands to moist, tropical lowlands. And so it is with many other crops, such as rice.

In India alone it is estimated that there are around 200,000 varieties of rice. This is a vast number, but it is only half as many as are believed to have existed in pre-industrial times. Much of this rice diversity has been lost recently because of the introduction of modern, commercial varieties that are better suited to modern intensive techniques. The same thing is happening to maize, and in some areas genetically modified varieties are cross-breeding with traditional ones, in the process further diminishing the traditional strains.

I cannot help but conclude that what was once a harmonious relationship between farmer and Nature is fast turning into an industrial process built on the flimsy foundations of exploitation, rather than what I would regard as the sounder footings of nurture and partnership. And this is of more than passing interest. The depletion of crop varieties is leading to the loss of different traits that could be of huge importance for people in the future – varieties that ripen at different times, for example, or ones that can withstand drought or are disease-resistant and that respond in different ways to manure. As we face many challenges, including climate change and the depletion of oil reserves, it might well be that traditional crop strains will once more underpin our food security – if they are still around to be deployed.

The suicide belt

The monocultures that now sprawl across vast areas of the world's farmland are based on plants that are effectively genetically identical. Some of these are modified to include genes from other species, which has not only led to the loss of wildlife but has also helped to create a terrifying loss of farmers. For a long time I have been particularly concerned about the situation in India, where small farmers are encouraged to borrow money at exorbitant rates of interest in order to buy genetically modified seeds, as well as fertilizers and pesticides, all of which are aggressively marketed by multinational companies. It appears, sadly, that, for whatever reason, the failure of harvests is a depressingly common occurrence and this compounds the debts the poorest farmers face. Having to buy yet more seeds that do not reproduce does not help the situation. The farmer has no way of paying back his debts and, of course, no crop either. The net result of such a situation is too awful to contemplate. In the past decade a staggering 100,000 Punjabi farmers have committed suicide because of the

Cows in a rotary milking parlour. While modern farming has increased yields, it has often treated animals more like machines than living beings. The manner in which farm animals are reared says something rather profound about how we have come to regard Nature. Once respected as a sacred gift, the natural world is now more often treated as a mechanism that we can test to destruction.

economic pressures that the industrial approach has imposed upon them. The Indian Parliament reckons that 16,000 suicides a year have been added to that total since the introduction of GM (genetically modified) crops. So, it seems we have devised an approach to farming that not only kills other plants and insects, it depletes the soil and indirectly kills the farmers too. Which begs the question: is this disconnected, mechanical approach to the task of food production really a long-term, sustainable path for the world to take?

Some would say wholeheartedly 'yes' – including the renowned architect of the 'Green Revolution' in India, Norman Borlaug, who won a Nobel Prize for his role in spreading intensive agricultural practices to poor countries. To call it a 'Green Revolution', however, is very misleading. Farms and fields were not being made greener. They were being turned into massive outdoor factories with a heavy emphasis on growing just one crop to the exclusion of all else. This kind of approach has wiped out techniques that had, for centuries, maintained a rich and vital biodiversity. Borlaug, though, believed that we must continue to use such a system because we face the pressing problem of how to feed more and more people. As he put it, 'without chemical fertilizers forget it.

The game is over.' But in light of the sort of evidence that is coming out of the so-called 'suicide belt' of India, it seems to me that *with* chemical fertilizers and all that goes with them, the game may well indeed be perilously close to really being over.

The hefty reliance on vast amounts of energy and chemicals, as well as the wider costs these cause to society, are generally excluded from calculations of the economic viability of intensive farming. The present way of accounting does not reflect sustainability questions, and this is why the diversity of food systems and the variety of plants and animals are replaced by a way of doing things that processes farm output into a variety of brands which, being packaged, give the impression of diversity and choice when in fact they are part of an underlying tendency to uniformity and monoculture.

Driven by official policies to promote 'cheap food', production subsidies, cut-throat competition in retail markets, readily available fossil fuels and ever more liberalized international markets (these last, in turn, driven by deliberate policy reforms aimed at the globalization of agriculture), food production and its retailing have been subject to a progressive process of intense industrialization so that today our food is produced, processed and sold by fewer and fewer huge companies. As a result, in many countries there are fewer farmers working land that has been dramatically transformed, mostly for the worse, so that its natural diversity has become depleted, its aesthetic appeal destroyed and the livelihoods it supports made all the more meagre.

Considering how little we know about the natural world, never mind what medical, nutritional, and other benefits it might provide for us, I have reached what seems to me, at least, the only logical conclusion: that it is the height of folly to continue with such a conscious destruction of what remains. If not for our own benefit, then surely we should act for the prospects of future generations by trying to stem the loss of unique and irreplaceable life forms?

But beyond even the consequences of eliminating ever more species of animal and plant, and beyond the continuing erosion of agricultural biodiversity, there is another level of natural variety that is perhaps of even more importance to us.

Utility

The term 'ecosystem services' has begun to be used more and more in environmental and conservation circles and is even beginning to penetrate the language of international treaties. It is a concept that has been central to some of my recent work, including efforts to encourage the world to see how vital

OVERLEAF:
Monteverde Cloud Forest, Costa Rica. Because of the almost constant cloud induced moist air, the trees are festooned with bromeliads, mosses and other epiphytes. This large reserve is widely regarded as one of the most outstanding wildlife sanctuaries in the Americas. It protects more than 400 species of birds, including 30 hummingbirds, and over 100 species of mammals.

the rainforests are to the very survival of many millions of people, which I will explain in more detail later.

Although perhaps rather an off-putting term, ecosystem services is one of great importance, for it succinctly sums up the wide range of benefits that people derive from properly functioning natural systems. In many respects the advantages provided by Nature are like giant utilities, and indeed include services that are instantly recognizable as such, for example the provision and purification of water.

A film sequence put together by NASA from satellite images and shown to me recently demonstrates the idea admirably. From a vantage point in space, it shows the annual pattern of cloud formation over rainforests. As the trees and other vegetation breathe and grow, so they exhale water vapour. Around twenty billion tonnes of water are released every day by the Amazon rainforests alone, and this condenses into great swirls of white cloud that then produce rain. Not only does the moisture fall back on to the forest, it also travels. Rain clouds generated by the Amazon rainforest help to water crops across a wide swath of South America, including the vast grain lands of Southern Brazil. Some of this moisture also falls over the grain fields of the prairies of North America. And these rainforests even help to water crops and replenish wetlands across North Africa and Southern Europe. One expert described to me how it is rather like a system of 'flying rivers' that move moisture around the skies, all driven ultimately by the rainforests themselves.

The statistics are truly staggering. For example, it has been estimated that an area of rainforest trees is able to evaporate into the atmosphere eight to ten times the amount of water that comes from an equivalent patch of ocean. Perhaps even more breathtaking is the fact that the energy needed to match the *daily* evapotranspiration (basically the release to the air of water from vegetation) of the Amazon basin's rainforests is about the same as would come from the world's largest hydroelectric dam working on full power for some 135 *years*.

Water is essential for our economic well-being, not only enabling crops to grow but also providing the lifeblood of cities and industry everywhere. And remember, in order to produce one kilo of beef it takes fifteen tonnes of water. Just one cup of coffee requires around 140 litres. It is a simple relationship: without the forests there is less water; with less water there is less food.

One of the major challenges that will confront humankind during the present century will be how to match our needs for fresh water with the limited, and in some cases, dwindling supplies that will be available. Urbanization, ever more intensive farming, the never-ending desire for more and more economic growth, and the inexorable increase in the global population will all lead to

greater demand at a time when climate change will be causing some areas to have less water. In these circumstances, making every effort we possibly can to maintain sources of rain, especially the rainforests, should be the top priority for the age we live in, and yet the destruction goes on.

And of course it is not only for water that we rely on Nature. Natural systems recycle our organic wastes; soil nutrients are replenished by an unimaginably vast army of microbes and fungi; we rely on insects for the pollination of crops; coral reefs and mangroves protect coastal areas from tidal surges and act as nurseries for the oceanic fisheries that help feed our cities; we rely on trees to help cool the air and the climatic conditions that sustain food production. There is also emerging science that tells us about the vast benefit gained from the storage of organic material in soils. This is not to mention what Nature does to nurture the human spirit, inspiring art and literature and embodying intrinsic values – although I have to say that in many societies today this is something that seems to lie increasingly near the limits of human comprehension.

We also tend to forget that for 3.5 billion years Nature has been working as a vast non-stop chemical laboratory. As a result, an incredible diversity of complex compounds has been produced. There is a tendency for chemists these days to spend more time in the laboratory than in the forests, in search of new potentially useful molecules, and yet there are many substances that because of

Sprinkler sprays water on crops, Palouse, Washington, USA. In addition to cheap oil and gas, modern farming relies on vast quantities of fresh water. Demand for water is rising fast, especially in agriculture. Food security is at risk because of potential water shortages.

their complexity will never be identified by experimental methods alone. For example, I recently learned about a species of frog discovered in Australia that had a particularly odd life cycle. Once they had laid their eggs and the male had fertilized them, the females swallowed the eggs so that the tadpoles might hatch and develop in the relative safety of the stomach. On reaching a certain stage these were expelled into the outside world to complete their development. When I learned about this particular means of reproduction I understood why these creatures had the rather odd name of gastric-brooding frogs.

In order to complete such a life cycle, various ingenious adaptations had evolved, including a chemical shield used by the tadpoles. When they were studied it was found that they secreted substances that inhibited the digestive processes and prevented the adult frog's stomach emptying. This capability is of great interest in designing future drugs and treatments for human peptic ulcer disease – a condition that causes misery to more than 25 million people in the USA alone. However, research could not continue because these frogs suddenly became extinct, in part as a result of the destruction of their forest and stream habitats. The miraculous chemicals that enabled the tadpoles to develop inside their mothers' stomachs and that might have taken millions of years to evolve – the chemicals that could have provided more effective peptic ulcer treatments – are now gone for ever. We will never know what they were or how they worked.

Marbled cone shell. The deadly poisons produced by this and other cone shells could provide major medical benefits, but only if we prevent the extinction of these remarkable creatures.

One group of animals that appears to be especially rich in potentially useful compounds is cone snails. These predatory creatures live on tropical reefs and in mangrove forests, mostly in the South Pacific region. There are about 700 different species and each is believed to manufacture 100–200 different peptide toxins to coat the paralyzing harpoons they use for hunting. Although only about six species and about 100 toxins out of a possible 140,000 have been studied in any detail, it seems that they offer the enormous potential of providing the basis of future painkillers and treatments for epilepsy. Some scientists believe that cone snails may contain more useful medical compounds for humans than any other group of creatures on Earth. And yet they, too, are under threat because coral reefs are being eroded by development, pollution and climate change, and also because mangrove forests are being cleared to make way for shrimp farms and other coastal developments.

What astonishes me is how all these arguments still seem to count for little in our ever more industrialized and urban societies. It seems that because of our mastery of science and technology we have convinced ourselves that we can somehow outflank Nature, to base all that we do on our technology and industry alone. Perhaps it is finally time to recognize what I have been saying now for a very long time: that changes to the climate and the destruction of Nature are not first and foremost environmental or 'green' issues: they are hard economic facts and matters that sit at the heart of how we ensure human well-being. For example, in relation to farming it should be a cause of great concern that we increasingly rely on ever more extreme forms of monoculture and have come to depend complacently upon a tiny handful of species to produce much of our food. In these circumstances it is evidently tempting for some to believe that we have successfully isolated ourselves from the need to conserve bio-diversity. After all, according to the United Nations' Food and Agriculture Organization, only some twelve plant species provide about three-quarters of our food supply and only fifteen mammal and bird species supply more than 90 per cent of global livestock production. However, these statistics exclude the fact that the productivity of these few species relies on hundreds of thousands of others – to recycle nutrients, to enable pollination, convert atmospheric nitrogen, to produce rainfall, control pests and facilitate the transfer of nutrients between soils and plants. Without all this, our handful of domesticated animals and plants would be useless. We must realize that our food is produced by a whole system, not just isolated elements. Again, the impression that we have somehow bypassed Nature is just that, an impression; and what I desperately hope the world will wake up to is that it is a very dangerous one. The evidence available provides us with ample warning of the likely consequences of con-tinuing to live out of balance with Nature, which is why I believe one of the most profound failures of our present way of thinking is seen in the realm of economics.

Eco-nomics

Almost every week my attention is drawn to a new report highlighting our dependence on one of the ecosystem services I have been describing, and to our continuing abuse of them. The recent collapse in bee populations in different parts of the world is a good example. These insects are vital for agriculture because they pollinate many of the plants that feed us, including beans and fruit trees. It is not clear yet why such a catastrophic decline is occurring, but, along-side disease epidemics, it seems that chemical pollution and intensive farming have possibly played an important role. It is perhaps difficult for us to see a bee as an essential worker in the functioning of the economy, but that is exactly what these insects are: components of economic stability and well-being. They are as important as any bank in the functioning of the financial world. These creatures, and millions of others too, maintain relationships and conduct transactions that are essential for our continued prosperity. Remove the bees and we are much poorer, perhaps even ruined – although, of course, there will be those who will say that some clever technology will be developed to take their place.

I would say the same thing about the rapid depletion of marine fisheries, a trend that has long concerned me and a subject that I have worked hard to find solutions to. The colourful selections of fish laid out in supermarkets are the fruits of complex ecological processes that are now being severely disrupted by pollution and over-fishing. Soon they may also be victims of the increases in absorption of damaging concentrations of carbon dioxide from the air, as I have hinted. The result is less fish available for us to eat.

The world's fisheries are worth $80–100 billion. They are the main or only source of animal protein for about one billion people and the process of catching them employs about 27 million people worldwide. In 2002 fish catches peaked and it is estimated that today around 70 per cent of fish stocks are being fished unsustainably. Industrial marine fishing has already reduced the total mass of large predatory fish, such as tuna and cod, to only 10 per cent of what it was forty or fifty years ago. And it is not only albatross populations that suffer as a consequence. Dolphins, sea turtles and habitats like coral reefs are all at risk from over-fishing. Sea beds across the world have been smashed up by bottom-trawling gear. Again, I have had the opportunity to speak with many experts on this subject, and they warn of impending disaster.

Even in the case of freshwater fisheries, which tend to be under better regulatory control, there is serious cause for concern. Pressures range from pollution to climate change and the way water is impounded using dams and diverted from rivers to irrigate farmland. And it is not only fisheries that are at risk from how we use and manage fresh water.

In 2006 I was struck by the findings of the United Nations Development Programme's Human Development Report on the way water is managed. This highlighted the disturbing fact that ancient, local, traditional approaches to water-harvesting are being rapidly abandoned as countries attempt to centralize and industrialize water resources. It described how, in many countries where new technology and new ideas are embraced in the rush towards modernization, the personal responsibility that individuals and entire communities once felt for maintaining their own water supply is also disappearing. This is interesting. It follows the general trend in modern thinking where rights become more important than responsibility. In the light of the failure of many large-scale modern attempts to centralize water management the report urged countries to think more about encouraging people to recognize their responsibility and to reinvigorate traditional approaches.

There are plenty of other examples of what I see as flaws in our appreciation of our fundamental *economic* reliance on Nature, and I have seen how these are repeatedly highlighted by scientific bodies, as well as an increasing number of economists and governments. International treaties to protect species and ecosystems have been agreed. National laws have been passed and some companies have begun to look at their supply chains, so as to better understand their reliance on natural systems. All this is very positive, yet tangible benefits for ecosystems are in many cases yet to be seen, at least on the trend-reversing scale needed. Part of the challenge lies in how we have conceived our economic system and how so-called 'market failures' can lead to devastating impacts on Nature that in turn harm the economy.

In recent years researchers have begun to look at the scale of this economic flaw and have reached some quite amazing conclusions. One study that made a big impression on me was completed in 1997. It was an investigation which set out to estimate how much, in financial terms, Nature is worth to us by calculating the cost of replacing the services it provides – if we possibly could. The seminal paper in question was produced by a research team led by Robert Costanza and was published in the leading journal *Nature*. It was called 'the value of the world's ecosystem services and natural capital', and summarized research that set out to estimate the value of a wide range of ecosystem services, including wetlands and how they protect property from flooding, the insects that pollinate crops, the benefits provided by rain, and natural regeneration of soils – among a range of other services. The figure they arrived at suggested the annual value of Nature in bald, bottom-line financial terms was then about $33 trillion. This, they said, was a minimum estimate.

What is perhaps more significant about this finding is how that figure was getting on for double global GDP at the time ($18 trillion). In other words, according to this calculation, the part of the economy that we measure, desperately try to grow and obsess about day after day in the Media is only about half as valuable financially as the part upon which we place almost no financial value at all and give little attention to, and yet which is the ultimate source of all our wealth! The term 'market failure' hardly does justice to the scale and profundity of this oversight, but that is what it is: an economic failure of epic proportions.

Honey bee pollinating flowers. These insects are not only a vital component in complex ecosystems, but also a vital part of the human economy. One recent study suggests that the retail value of agricultural products produced in the UK with the help of honey bees pollinating flowers is about £1 billion per year. In parts of China where honey bees have disappeared farmers must pollinate fruit tree flowers by hand with feather dusters.

Other studies only serve to clarify this state of affairs. One, begun in 2008 by the United Nations and called the Economics of Ecosystems and Biodiversity Study, or TEEB, which continues to do its work, sought to highlight the growing cost in financial terms of the annual loss of biodiversity and the destruction of ecosystems around the world. They began by working out the value of the natural systems wiped out every year, be it wetlands or rainforest and so on. If you imagine this area as a factory, they estimated that this factory would be worth around $50 billion. But that is not the total loss. The figures are worse than that. Calculating the annual output lost from this $50 billion economic concern over a period of the next forty years and applying a low rate of interest to that output, their estimate was that, in financial terms, the world's economy is incurring a loss of between $2 and $4.5 trillion every year because of our destruction – every single year. To put that figure into perspective, when the world's banking system suffered its crash recently – described by the Media frenzy as 'the worst financial crisis the world has seen since the 1930s' – the estimated one-off loss of that crisis was put at just $2 trillion. We all witnessed the scale of that news story, but it is curious that the far bigger one has never yet received the same sort of panicked Media attention.

This leads me to suggest that the increasingly unbalanced relationships we forge with the world around us are of far more importance than is widely understood. Harmony between people and the rest of Nature is not simply a philosophical or ethical matter – it is a fundamental practical and economic priority. But it is not seen this way because of the way we have come to measure and consider economic progress.

During the twentieth century, and particularly in the period after the Second World War, countries began to measure something called Gross Domestic Product, or GDP. This is basically a proxy for how much economic activity is taking place in a country and is a measure of the output of goods and services. Once it became possible to measure GDP with some accuracy (and also Gross National Product, or GNP, the measure of income rather than output), it was a natural and short step to measuring 'growth' – the increase year on year of GDP or GNP. This proved a very convenient tool for assessing a country's success, not least because it gave the impression of being a good benchmark for social welfare and national progress.

I grew up in this era and have seen the vast changes that have occurred in society as a result of what has become our obsession with endless economic expansion. I can see that measuring growth in this way is the logical outcome of two and a half centuries of industrial development in the West. It is, after all, a means of computing our continued and accelerating ability to improve human progress by harnessing technology and pushing the boundaries of science. What is perhaps less well known is that those who came up with the idea of measuring GDP growth cautioned against using it in this way, simply because this is not what it was designed for. Economists such as Simon Kuznets, who helped to establish ways of measuring GDP, warned that there was a lot more that it didn't measure, compared to what it did. He wrote that "the welfare of a nation can scarcely be inferred from a measure of national income". GDP nevertheless offered to begin with a broadly successful means of helping to create relatively orderly ways of expanding wealth and helping people gain access to more goods and services. One other outcome was that, as people got richer, so the flow of taxes to governments got bigger. As time has gone by, though, it has become increasingly worrying that the limitations of GDP growth are not being reflected in how economic development is pursued in many countries, although at least now the limitations are becoming much clearer.

One problem with GDP growth as a central measure of progress is that it only measures certain things. Others are left out of the equation. For example, much of the welfare enjoyed by societies derives from the quality of people's relationships and the pleasantness and security of their neighbourhoods. GDP does not measure change in these. Neither can it accommodate the financially unmeasured but very real benefits that derive from good parenting, or the care received by an elderly or infirm person from their families. It does not measure how happy we are, nor whether our lives are fulfilling.

Neither does GDP reflect the huge costs that come with clearing ancient forests, depleting fisheries, or loading carbon dioxide into the Earth's atmosphere. Worse still, all these are the result of activities that at the moment *increase* economic growth. The clear-up of a major pollution incident contributes to growth; so does the sale of the complex drugs needed to treat our twenty-first-century health problems like cancer, heart disease and widespread allergies. While all of these things count positively towards GDP growth, they are at the same time either signs of diminished natural capacity or reduced human welfare. This is why I think there is now a very strong case to conclude that we are measuring the wrong things. The picture is incomplete.

We have inadvertently created economic signals and measures that regard many natural forms of capital as valueless, not least the stability of the climate. This seems to me to be a fundamental and pretty remarkable oversight, considering how the connections between the continuing degradation of Nature and its economic impact now stare us in the face. For example, some 75 per cent of the electricity produced in Brazil comes from large hydro-power dams. They are clearly totally reliant upon rain which, in the main, is produced by the rainforests of the Amazon basin. Yet as the forests have been cleared in pursuit of economic growth, and the 'benefits' of deforestation increasingly judged from this perspective, the *cost* of clearance, for example, has not been factored into the future price of producing electricity. In other words, the short-term value of deforestation is not set against the slightly longer-term rises in the price of power that it will cause, never mind the myriad other benefits that are being lost.

Unfortunately, the blunt truth of the situation at the moment is that in order for people to contribute most to national success – that is, success measured by the growth in GDP – they might drive everywhere in a huge, energy-wasting car and then buy a new one every year. They might also buy vast quantities of unnecessary consumer goods, waste much of their food rather than eat it, and, after retirement, die a lingering death preceded by years of dependence on expensive, life-extending drugs; all of which would contribute positively to GDP growth. It may sound a pretty miserable approach, but these are the kinds of things that maximize our principal measure of economic success. With this as the dominant mindset it is perhaps little wonder that such grave imbalances occur.

More recent studies confirm that this economic contradiction not only continues but deepens. For example, the 2006 UN Millennium Ecosystem Assessment demonstrated how it will be impossible to meet our long-term aim of reducing poverty if we continue to deplete and destroy the Earth's natural ecosystem services. Later that year the Stern Review on climate change was published. It set out how, if we do not act very soon, the ongoing pollution of the atmosphere can be expected to cause damage to the economy in the coming few decades that would be equivalent to the cost of both world wars and the Great Depression combined. To compound the urgency of the situation, since the publication of his report in 2006 the author has suggested that his original projections were based on an underestimate of the speed and severity of climate change and its likely costs.

When GDP was first adopted as the main measure of economic success the assumption was that growth would make us happier because it would provide the resources needed to make us more comfortable and that would offer us more options and greater freedom. Up to a point this has been the case. But

The largest power station on Earth – Itaipu hydroelectric facility, on the River Parana between Brazil and Paraguay. Each one of the massive pipes seen in this picture is 30 feet wide. Much of the rain that flows in the river which powers this huge dam is generated by the Amazonian rainforest.

even this logic is now under attack. It seems that beyond a certain stage the relationship between increased growth and happiness starts to break down – and that point now appears to have been reached in many parts of the world, especially the richer ones. Academic research has found that consumerism with its emphasis on the acquisition of more and more material goods does not for many people lead to increased happiness. It seems it does not nurture increased well-being once a certain level of comfort and security has been attained. As we shall see in the last chapter, this is all part of the design of consumerism, which again is a more structured ideology than you might imagine. Built into it is both a means of stimulating a desire for happiness by having, and an assurance that this desire will never be completely satisfied.

Professor Tim Jackson of the University of Surrey has called the challenge we face the 'dilemma of growth' and recently put his name to an open letter to Her Majesty the Queen that described the underlying cause of the recent economic meltdown as 'a multi-generational debt-binge, inextricably linked to a concomitant multi-generational energy-binge'. He is not the only one to point out that bingeing does not lead to happiness. The Nuffield Centre in the UK has been conducting ground-breaking research on adolescence with a meticulous 'time-trend' study that began in 1974 and now spans three decades. It reports that the country's young people now have 'significantly higher levels of emotional and behavioural problems than 16-year-olds who lived through the 1970s and '80s'. Clearly, as we continue to liquidate the world's natural assets in pursuit of what we call 'progress', the many social challenges that we hoped economic growth would solve – poverty, stress and ill health – for example seem reluctant to respond to the cure of yet more consumption.

Mahatma Gandhi made a crucial observation when he said that humanity has a natural tendency to consume. The crucial element that he felt was missing from much of the Western approach to life was that of limit. If there are no limits on that tendency, he explained, we can become obsessed with satisfying our desires, consuming ever more as we chase what little satisfaction we achieve. Gandhi was also very clear about the danger of this tendency if it is legitimized by a view of the world that puts humanity at the centre of things, operating under the assumption of an absolute right over Nature. He predicted that such a combination would prove explosive. It would be, he declared, a very destructive world view indeed.

I think the evidence suggests that Gandhi was right, and not just from an ecological point of view. Many developed countries have reported long-term increases in mental health problems. The combination of the stress of trying to keep pace with rampant consumerism and the impact of people living more isolated lives has led to many millions becoming victims rather than the beneficiaries of how we have chosen to achieve and measure progress. Our physical health has also suffered. Many millions of people are now classified as clinically obese because of their much more sedentary lifestyles and their fat-ridden diets. It is a depressing portrait of our age that we recently reached the point where the number of obese people in the world surpassed the 800 million who are estimated to be malnourished. This is but one awful example of how we are now living in a world of widening extremes. Carrying on as we are, ignoring the need for balance and a more integrative form of economics, is only going to force that gap to grow wider.

As we continue to liquidate the world's natural assets in pursuit of what we call 'progress', the many social challenges that we hoped economic growth would solve – poverty, stress and ill health for example seem reluctant to respond to the cure.

I am only too aware of the argument from developing countries that they should not be denied the benefits that the developed world has enjoyed, albeit to such an overblown degree, but we must all recognize what will happen if we do not think again and think constructively about how to build a better economic system for the future, one that understands the limitations and dangers inherent in our present industrial mindset. We have to find ways of ending poverty, but we also need to look at the way societies have developed in the richer parts of the world. We need to question the unbridled encouragement of consumerism and, I am afraid to say, we also have to address that issue that so often is side-stepped as being just too hot to handle, the question of population increase. Not only because of what will happen to the very life-support systems of our planet if we do not do so, but also the consequences this will have on the welfare of people. As the biologist Paul Ehrlich has pointed out, 'there is no technological fix that will allow perpetual population and economic growth.' At some point we will need to recognize that there are very important limits to what Nature can withstand, and that these limits must

determine what we can demand of the stressed systems we rely upon. I am absolutely sure that this question of population growth has to be part of the debate about developing a different philosophy for living.

After all, the simple fact that I have tried to demonstrate so far is that pursuing ever more conventional economic development, based on growing the economy by promoting more consumption of goods and services, and doing so for billions more people in the next forty years, will place an impossible strain on the finite resources and inherent capacity of the Earth to renew and replenish herself. The simple arithmetic says that we cannot expect to succeed.

If, as some economists have done, we consider the services we derive from Nature as if they amounted to an annual income, then, as an example, in 2008 we had used up our entire yearly budget by mid-September. That was a few days earlier than the one in 2007. In both cases, from then until New Year's Day we were living by liquidating our capital assets: the forests, soil, fresh water, fisheries and biodiversity. So we are already operating on a diminished return, and that is with 6.8 billion people. If the world's population continues to balloon as every prediction says it will, and if economic development continues at the pace we are forcing it to, then this 'credit overdraft' is set to get a lot bigger and its effects a lot worse. By mid-century the idea of making it to September will be a pipedream. We will have used up the Earth's depleted services by April and by then the degradation of our capital assets will have put Nature close to going bust.

By measuring how rich we are, and in effect how many resources we are using up, and seeing growth in both of those measures as wholly positive, we operate according to a very partial view of reality. The writer and philosopher Satish Kumar sees Western societies as having become dangerously confused by equating money directly with wealth. He points out that 'money is not wealth; money is only a measure of wealth and a means of exchange. Real wealth is good land, pristine forests, clean rivers, healthy animals, vibrant communities, nourishing food and human creativity. But the money managers have turned land, forests, rivers, animals and human creativity into commodities to be bought and sold. Even money itself has become a commodity as speculators trade in money to make more money.' This has gone on for some time, of course, but has become ever more extreme, with this last step leading to an economy that is not only increasingly divorced from the realities of Nature, but that has undermined even its own narrowly conceived aim of 'growth' through the recent financial collapse and then a difficult recession.

This is surely the business model of the madhouse and realizing the folly of it should really prompt us to raise the question of whether it is possible to

LEFT: *Aerial view of crowded favela housing contrasts with modern apartment buildings in São Paulo, Brazil. Scenes like this remind us that rapid economic growth does not automatically solve pressing social challenges. Despite decades of growth many countries remain socially divided at the same time as environmental damage has accelerated.*

sustain the idea of 'capitalism' as we have come to know it if we end up without any capital: that is to say, without Nature's capital. And we should not be under any illusions here. Unlike those crises in our financial system, this failure of the economy will not be so easy to fix. We will not be able to pump more cash into the system. Much of the ecological capital that we are burning up at the moment is irreplaceable: as we have already seen with the gastric-brooding frogs, once it has gone we cannot get it back again.

I want to show how this view remained a constant through many centuries and in many cultures – indeed, how it created the foundations of Western civilization and supported it right up to that febrile period in European history I shall soon describe.

I feel it is now time to face the undeniable conclusion that we are currently on course for a massive and rapid ecological decline. It is as if we are sailing on a giant tanker and heading straight for a hurricane. I know well from my time in the Royal Navy that if it was simply a big storm we would be able to sail through it – not a pleasant experience but possible. But at sea you do not sail into hurricanes. You go round them, respecting their immense power to destroy. We can see this danger ahead and we need to change our course, but tankers do not turn swiftly. It will take a monumental effort if we hope to avoid disaster – with all hands to the pumps and all engines brought to bear. We do still have time to turn the tanker, but not much. What follows may at first appear to be rather irrelevant to the pressing problems I have just described. But I hope it will become clear that, far from having no bearing on how we face these many challenges, looking at how people in the past viewed the world offers us an important way of seeing how we now regard the world, our place within it and perhaps how some of the flaws have emerged in our perception.

Throughout all of my attempts to find and test possible solutions, I have become increasingly certain about one challenging fact: that the reason we have ended up in this mess and why we continue to dig ourselves ever deeper into it, despite knowing all along that we are doing this, comes down to the way in which we have been persuaded to think and how we perceive the world. What began as a laudable approach has gone way too far. I am sure that those pioneers of the Scientific Revolution in the seventeenth century and the fathers of the Enlightenment in the eighteenth would be horrified at the scale of things today. I imagine they would be reluctant to condone what has happened in their name, for they never intended that the world be subjected to the sort of giant experiment we now conduct, not just on the grain-giving soil but on the way we design and build our houses, workplaces, towns and cities, on the way we view medicine and healthcare, the way we view communities and, indeed, on the view we have of our own sense of being.

This is why I now want to explore the journey that has produced this peculiarly aberrant world view. I want to take you back in time to reveal a less well-known aspect of the way ancient civilizations viewed their place in the world. I want to show how this view remained a constant through many centuries and in many cultures – indeed, how it created the foundations of Western civilization and supported it right up to that febrile period in European history I shall soon describe. This is not to suggest that we should be blind to the fact that ancient societies and their civilizations were, by our modern standards, cruel and inhuman, or that they were ravaged by the sorts of plagues and diseases that modern science has either eradicated or found cures for; that life expectancy was short, the food was often poor or that healthcare was often non-existent. But by looking at their deeply rooted perception of the world and their place in it, expressed so eloquently in their sacred art and the symbolism embedded in their sacred architecture, it is possible to see how, as a mechanistic and secular view of the world grew from around the seventeenth century onwards, the seeds that sprouted at that time produced the legacy of our current and increasingly destructive view of the world.

What I want to show is just how fundamentally at odds with reality our modern view has become compared with the one that sustained the world for thousands of years. And I mean 'fundamentally' because, as I hope will be very clear when we take a tour of the geometry of the natural world and how that patterning was reflected in all of the great sacred architecture of the past, the way we look at the world now is not just one that is somehow *different* from an older, alternative view. It is fundamentally at odds with the way in which the Earth behaves and even contrary to the way in which the entire universe operates. It is why I cannot stress the point enough: we are travelling along a very wrong road.

The Golden Thread   3

From harmony, from heavenly harmony
This universal frame began:
From harmony to harmony
Through all the compass of the notes it ran,
The diapason closing full in Man.

JOHN DRYDEN

The high points of human civilization have all been framed and shaped by what I have come to see as 'shared insights'. These insights belong to humanity as a whole. They are not the preserve of one tradition or one school of thought, nor do they come from a particular moment in time. They are timeless and universal and the wisdom they convey is embedded in all of the world's great sacred art and architecture. Such art and buildings are concrete expressions of this traditional perspective.

I happen to believe that these insights are tremendously important to humanity, but my contention here is that our modern approach has lost sight of them, and in so doing the Westernized view that now dominates so much of the world has become disconnected from its important anchors. I want to explore these insights and demonstrate what it was that people in the past felt anchored to, because they offer a perspective that could help us re-frame our approach for the future. It is a big subject and we will cover many hundreds if not thousands of years in the next few pages, but the journey should be swift. It is certainly necessary because it gives my proposition here the context it needs for it to be properly understood.

As I have already made clear, at the heart of the matter lies a crisis in our perception – the way we see and understand how the world works. Grasping the causes of this ultimate crisis is a first and important step if we want to find solutions to the problems I have just explained. For we have to understand what is missing from our present picture of the world in order that we can put it back again so that any solutions are well rooted and work in the long term.

This chapter will begin back in the mists of time, but it will end up at that point in our collective history when the modern world was born. That

PREVIOUS SPREAD: *The Egyptian goddess Ma'at, tomb of Queen Nefertari 1300-1255 BC. Ancient Egypt harboured a profound wisdom. The Egyptians considered the realm of myth to be where spiritual forces assume objective shapes to teach humanity universal realities.*

LEFT: *The ninth century Buddhist Monument, Borobudur, Indonesia. The entire monument charts the pilgrim's path from the lower world of desire to the highest one of formlessness and every element of it conforms precisely to the geometry described by Plato and found in many of the world's sacred structures.*

happened amid the complex turmoil in Europe in the seventeenth and eighteenth centuries. Such was the impact of the events that took place at that time, the advent of the Scientific Revolution in the seventeenth century and the emergence of the Age of Reason in the eighteenth century, that our understanding today of the way the world was seen before that time is somewhat obscured. It is as if all that happened from the seventeenth century onwards built a very high wall so that only the tops of the trees are visible in the landscape beyond it. I want to try and remove that wall for a moment and see how the world looked without it – simply because I believe that this perspective reveals a view of reality that could be incredibly useful as we struggle to wean ourselves off our excessive dependence on the mechanistic approach to life that is now so embedded in our culture.

To demonstrate this traditional perspective as it passed from one period of civilization to the next I have arranged this chapter into three parts. In the first I want to consider the most important elements of what I have called 'the grammar of harmony' as it was understood by ancient civilizations. It framed their entire understanding of life. This is necessary in order to provide a wide enough context for what the second part more specifically deals with, which is a graphic illustration of how the grammar of harmony really works, how the language of patterns that are found throughout Nature fit together and hold the very fabric of the material world together. In my experience, once somebody sees this in action, particularly the extraordinary correspondences that exist between the patterns made by the orbiting planets and the forms found in Nature, they tend to grasp the central point I have been trying to make for so long, which is, essentially, that the art and architecture that flowed from this geometry are not merely clever. There is a profound symbolism at work that is as relevant today as it has always been. And this is why the third part explores those later spectacular examples of sacred art and architecture as they came to life during two more recent high points in civilization, one Islamic, one medieval Christian, both of which sought to create a three-dimensional expression of the profound insight their rediscovery of the grammar of harmony had attained.

Each piece of sacred art or architecture we will explore was created according to the timeless principles of the grammar of harmony. They are all outward monuments to their times and to the religious traditions in whose name they were made, but in each case it is those timeless principles of harmony, balance and unity that concern me here. That is why this is about more than only history. I want to unravel a portrait of these that lie at the very heart of life itself and give shape to things. They are all too easily forgotten in our technically sophisticated, totally mechanized world. Every culture of the past has understood their importance and has used them to underpin the structure of their most important sacred buildings and many secular ones too. These principles also inform their religious symbolism and open up a clear experience of a deeply anchored view of the cosmos and of humanity's spiritual role within creation. It is this traditional insight that I firmly believe could be of such help in our own troubled and philosophically impoverished times. If that seems fanciful, then I would urge you to read on. The meaning that the symbolism of these buildings conveys is far more than information, but to understand it we have to experience it – to know it as a comprehensive scheme.

Art and architecture, music and poetry are the means of doing this because they come from the heart rather than from the head, so this is why I have chosen to explain what I mean by 'the grammar of harmony' in as pictorial a way as possible and by looking at the monuments and the cultural icons these great periods of civilization left behind.

Sacred geometry

Every great sacred building possesses an enigmatic beauty and the power to move us. It does this whether or not we happen to come from the culture that built it. Somehow the architecture strikes a chord in our hearts. We do not find it 'clever'. It does not shock us or 'challenge our assumptions', as contemporary art sets out to do. It resonates and somehow harmonizes with our own human nature so that we feel something is 'meant' by it.

I have always found it remarkable that the same shapes convey the same meaning, even though they are found in buildings that may lie thousands of miles apart and their construction is separated by many hundreds or thousands of years. Be it the great shrine to the Lord Buddha at Borobudur in Indonesia or the Parthenon six thousand miles away in Greece, or one of the great Gothic Cathedrals of Northern Europe, the same principles apply. What I want to show is how they give the building its presence as well as assuring its stability, which makes the geometry not just beautiful art and a means of expressing the deepest spiritual symbolism, but also a demonstration of ancient excellence in engineering.

The patterning that forms the geometry that we will explore in the second part is derived from a very close observation of Nature. The great architects and artists of the past did not study Nature's patterns simply because they found them pretty; they knew them to be the very patterns of life. They are just as innate in us as they are in a tree, so we need to be clear from the outset that, by

The artesonado ceiling, Mezquita Mosque, Córdoba, Spain. These intricately carved, inlaid ceilings are Moorish in origin but are also found in Christian churches and early Spanish synagogues.

studying the interconnected relationship between growth and order in the universe, the ancients were also exploring what lies at the very core of life – the element that makes it sacred. The wonderful architecture they created using this geometric code, and the many intricate pieces of very beautiful sacred art that were inspired by their contemplation, were an essential part of the process of anchoring humanity to the greater presence in life.

The journey that the transmission of these shared insights takes through human history twists and turns like a golden thread. It has never been immune to the conflicts the world has seen, so it is often to be found weaving precariously alongside the many horrors and atrocities that fill our history books. At some periods in history it falls into decline in one culture only to be kept alive in another. There are even moments when this golden thread of ancient understanding is to be found flourishing simultaneously in two cultures that might be at war with each other, which has made me wonder if perhaps we might not call it a separate history of civilization altogether, one that has little concern for cultural or national boundaries or even the politics of the day. Its priority, wherever it appears, is for what Coleridge somewhere calls 'the politics of eternity'. Its purpose is to maintain meaning rather than wreck it.

Seen this way the history of civilization is much more interconnected than the history books suggest and so it is not odd to conclude that we have the ancient Greeks to thank for the Islamic patterning that adorns every great mosque from Córdoba to Delhi, just as we have Islamic culture to thank for the precise geometry of every Gothic cathedral, from the South of France to York, whose towers soar into the sky. But let me begin this colourful tour with the one civilization that pre-dates even the Greeks. It is a civilization we all have so much to be grateful for.

Ancient insight

The culture and thinking of ancient Egypt that flourished 5,000 years ago could be called a foundation culture. Many aspects of our culture have their deepest roots in the land that clings to the banks of the River Nile. From that thin strip of fertility in North Africa arose much more than reeds and wheat. It is a source of our mythology and religious symbolism, our astronomy, geometry and mathematics, even the shape of many of our letters. They are all distant echoes of an outlook that defined a people whose life revolved completely around the ebb and flow of the mighty Nile.

Viewed from a satellite, Egypt resembles the cone of a tornado. It twists like a crooked dagger out of the sands of Sudan until it bulges into the great green, fertile cloud of the Nile Delta where it meets the Mediterranean. At points on the river the strip of fertile land is so thin that the white sands of the desert are only a stone's throw from the water's edge, so ancient Egypt would have been a culture shaped by a knife-edge dependency upon a very thin strip of fertile soil. Egyptians would surely have been haunted by the wilderness and extinction that lay beyond their fertile land and they would have been in no doubt that their lives hinged entirely on the benevolence of the river.

Once every year it would flood the plain to swamp the land with a vital cloud of black soil. The Egyptians called this black gold *khem*; this is where we get the word 'alchemy' from, and our modern word 'chemistry', and the meaning has little changed. *Khem* was a magical substance without which not a single thing would grow. It was lifted from the delta of the Nile as the waters rose during the time of the annual flood. With its capacity for abundant creation as

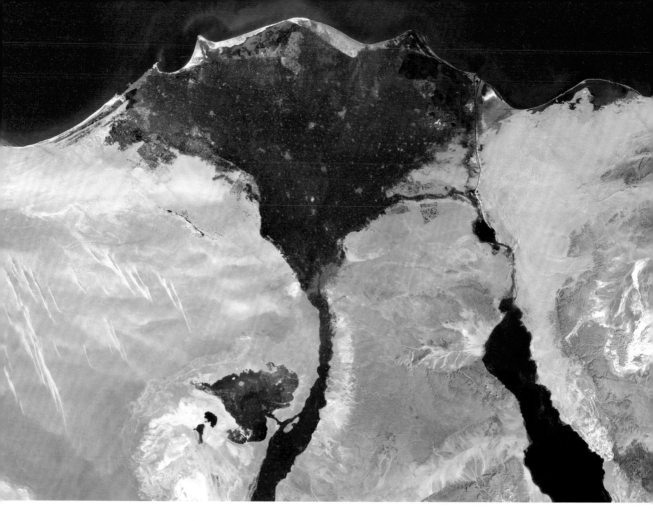

The Nile's delta clearly visible against the wilderness of the desert and forming the shape of the Ankh, the symbolic 'Key of Life'. The Egyptians considered their land the 'temple of the entire cosmos'.

well as desolation and destruction, the river did not just bond the ancient Egyptians to its cycles from a practical point of view; it also framed their outlook and their imagination. The river's cycle was woven into their mythology so that their gods became symbols of the perennial struggle between the harsh forces of decay and those of benevolent renewal. The stories that arose from this process are so primordial that we still live with strong traces of them today. Just think of the Egyptian god Osiris, the hero sacrificed in a brutal execution at the turning of the year, who then miraculously rose from the dead to be put back together so that he could redeem the world with life's vital force.

Such 'myths' are not fanciful stories. We would not be so drawn to them if they were. It may surprise you how often modern-day stories conform to their patterns. The characters and costumes may have changed, but millions still flock to buy them whether they be retold in the books of CS Lewis or JRR Tolkien, whose stories have now been made into box-office cinema hits or, more recently, the stories of JK Rowling or Philip Pullman that have captured not just children's imaginations, but their parents' too. All of these authors cleverly make the universal mythological patterns accessible in a contemporary way. What is less to the fore, though, is any intelligible explanation in simple terms as to what these universal myths actually mean and how they relate to our future survival. For that is the point of them. It is through their transmission, not just from one generation to the next, but from one culture to another, that the essential patterning in life is understood. It is a patterning that comes from a careful study of how Nature's balance depends upon the limits that 'contain' her unity and maintain her coherence. Coherence is another way of describing harmony and the Egyptians understood how harmony worked. So much of the symbolism in their art and architecture demonstrates that they held harmony to be supreme and a vital state. So much so that their most important deity was considered the goddess of harmony.

Ma'at

This goddess was Ma'at, who would judge the record of the Pharaoh at his death. She would weigh his soul against the ostrich feather she carried in her crown to decide if it could proceed to the afterlife. If the scales balanced, all was well, but if his misdeeds tipped the balance, his soul would be instantly condemned. In this way Ma'at was also the goddess of truth and represented the rightness of the law and of universal order. In her name were these things judged. She was said to be present at all moments of justice, just as today it is the Stars and Stripes or an image of the British Crown that sits above proceedings in courtrooms to represent the presence of sovereign truth.

The Egyptians are well known for their worship of the Sun god, Ra, but it was Ma'at who was supremely important in their imagination and, although thousands of years have passed since the Pharaohs were laid to rest in their tombs, the qualities that Ma'at represented remain just as important to us today. Not only was she the goddess of truth, the Egyptians believed that the whole world was maintained by Ma'at's active presence. Without her, the entire universe would fragment and collapse into the primordial chaos from which it had come. Ma'at was the very essence of harmony and it was therefore the primary duty of every Pharaoh – 'the Beloved of Ma'at' as he was called – to safeguard her presence. He maintained the law and administered its justice to ensure that harmony prevailed between Heaven and Earth. To do this the Pharaoh had to be both a priest and a king – a priest on the inside and a king on the outside – someone who had attained complete spiritual integration and had reached his full potential so that he could lead others towards the same.

'Ma'at' means 'plinth' and it is not by chance that the image of every other

god in the Egyptian pantheon always depicts them either sitting, kneeling, or standing on a plinth. It is a very explicit message: all of the gods – that is to say, all of the various qualities that make up the ordered universe – depend completely for their stability upon Ma'at. Harmony must always be present for them to exist.

This was not some mad idea dreamt up by a priest who had been sitting for too long in the sun. The scholars of ancient Egypt came to this conclusion for very good reasons. To understand how they arrived at such a perspective it is important to appreciate how vital their symbolism was to their culture. Having spoken with some of the most eminent Egyptian scholars of our day, it is clear that the tomb paintings and papyrus texts they left behind were created by a people alive to their meaning in a way that is hard for us to imagine. As they mapped out those murals in black and red inks the artists who created them attached symbolic importance to absolutely every stroke they painted. Their work was 'sacred' art. It was a form of prayer, a vision of the universe conveyed in a pictorial language that did not depict the 'outside world'. It mapped their experience of the inner realm, which was supremely important to ancient civilizations, to the extent that they considered the inner world the very ground of reality. It was the source that informed the 'lower' realm of form and matter

The goddess Ma'at (left), daughter of the Sun god Atum-Ra and the feminine counterpart to Thoth, the divine mind whom the Greeks called the Logos – the Word that begets Creation. Here she offers protection to Serket, the deity who guards against poisons.

and where everything begins. To depict this inner, or higher, realm was to activate the soul and to draw down the qualities of this higher, inner reality. In other words they were bringing alive the realm of spiritual reality in order to empower the lower, mundane world of space and time and things with spiritual significance. This has always been the purpose of sacred art and architecture. It is the process of 'earthing' Heaven and is just as alive today as it was 3,500 years ago, for instance in the world of artists who paint icons for a Greek or Russian Orthodox church.

I happen to believe that this is the underlying meaning of that much-maligned word 'superstition'. Far from concerning itself with 'hocus pocus', I suspect it once referred to that sense of something 'super' to, or above, the ordinary world that is difficult to articulate in words, but easier to know through the experience of sacred art and architecture. Hence my reason for exploring this lost perspective through some of the greatest works of art and architecture left to us by earlier civilizations.

Tradition

This is the basis of the wisdom transmitted by what is known as 'tradition', which is another word that, in the swirl of our 'anything goes' post-modern confusion, has lost its meaning. These days 'tradition' tends to refer to things that are 'old-fashioned' or backward-looking, when in fact the true definition could not be further from that. Tradition is a living presence. It looks to the future as much as it does to the past because it is focussed on the non-material as well our material needs of the day, neither of which have changed, despite our many advances and supposed progress away from all those 'old-fashioned superstitious beliefs'. We are still human after all. We still have to face the same central paradoxes of life and we still need a deep keel to our boat – not to mention a compass that will navigate us safely through the many storms and trials we have to experience.

What we can learn from the way ancient civilizations like the Egyptians looked at life is how they saw the same shape to things – the essential, cyclical process of growth that is limited by the need for decay, which in turn renews again into another cycle of creation. This is the pattern of Nature which, incidentally, I cannot help feeling gives a deeper meaning to that other word, 're-creation'. Whether we choose to acknowledge it or not, we are still bonded to this essential organic pattern. Only the seductive allure of the level of materialism that now drives modern consumerism has distorted this perception of the way the world works. We can end up forgetting what the ancient

Egyptians knew well – that we need to exercise a necessary degree of self-discipline and self-limit for the cycle to proceed unhindered. If that did not happen in their world, they believed they would unleash a force that could sweep everything away.

The force of chaos

They called this force *Isfet*. It was not a deity and nor did they ever depict it in their temple art or in papyrus texts, yet it lurked in the shadows of the world at large and in the dark recesses of their minds, threatening the stability of everything. *Isfet* could swamp their world with chaos and, as it did so, it would also infest the human heart to bring about unbridled social collapse. I mention this simply because, surely, the parallels with our own time are all too clear. Ignoring the threat of *Isfet* still puts us on course for catastrophe.

This is not to say that because they kept *Isfet* at bay and Ma'at happy that life in ancient Egypt was somehow perfect and miraculously without conflict, crime, poverty or suffering. Of course it was not. The history books are full of gruesome evidence of appalling massacres during that time. In fact there is enough suffering in every chapter of world history to dismiss immediately the romantic idea that a certain historical period could ever be called a 'Golden Age'. Life was harsh and short. Disease, starvation or invasion could wipe out entire societies. And yet there they are, these ancient artists and scholars, deeply alert to the foundation principles of harmony that have become so neglected in our own age with such devastating results. Eventually the Egyptian civilization sank into the sand, but not its wisdom which, in time, found itself forming the basis of a world view promoted by one of the founding fathers of Western civilization.

We know very little about the life of Pythagoras. He was born on the tiny island of Samos, one of the most fertile of all the Greek islands in the Aegean Sea. He supposedly travelled to Egypt in the sixth century BC and stayed there for twenty-two years, studying with and being initiated by the Egyptian priestly scholars. When he returned to his island home it is said he brought a remarkable

A detail of the facade of the new library at Alexandria, built to commemorate the great library lost in antiquity.

understanding of mathematics and a profound philosophical insight, but he did not stay long. Samos was ruled by a tyrant whose intolerance forced Pythagoras to go in search of a better home and so it was that he eventually settled in Croton, the present-day Italian city of Crotone, in Calabria, where he set about creating political reforms and in effect establishing a religious teaching community.

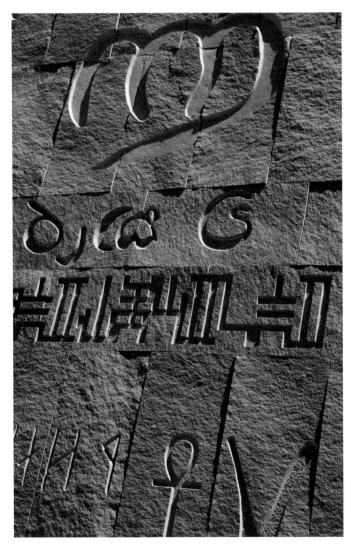

Pythagoras's teaching was based upon the essential kinship of all living things. He is said to have insisted that his followers should be, like him, vegetarians. His disciplined approach to diet and nutrition surely qualify him to be one of the earliest known dieticians because he linked healthy eating with a healthy mind, body and spirit. However, it is the central cosmic importance he gave to number which has always fascinated me – a scheme which is not difficult to understand, but one which states quite clearly that we live in an integrated and harmonious universe.

Pythagoras taught that there is a

precise relationship between the arithmetic of number and the geometry of the physical space around us. In his view, chaos is ordered by number and Nature is made up of precise numerical patterns. For Pythagoras, that meant the nature of number went far beyond being simply a means of calculating quantity. He held that the nearest the human mind could get to the Divine Mind was through number and, thereby, the principles of proportion and harmony. In this way we should guard against treating the Pythagorean teachings as merely technical matters. They give insight into both the practical world and the immortality of the soul. Number had a living, qualitative value and was symbolic of the higher realms of reality, those levels of reality that lie beyond the touchable, 'actual' world. It may seem an odd thing to say in this day and age, but Pythagoras taught that number expressed a divine quality. Not only is the natural world constructed according to a precise mathematics, Pythagoras suggested that if we contemplate its patterns deeply we are led into communion with the very source of number itself, which is unity. This makes the study of number as much a meditative process as a 'scientific' one – which is an aspect of mathematics the modern maths class certainly does not explore.

Before we look specifically at how the grammar of harmony works and then go on to see it in action in so many splendid ways once it was rediscovered 1,000 years after Pythagoras, it is just worth reflecting on what he meant by these patterns leading to a deeper meaning. It will make the symbolism of the geometry we are about to immerse ourselves in all the more comprehensible. Pythagoras considered that the first number represents the ultimate principle of unity. The universal symbol of unity is not the numerical figure 1. It is the sphere or the circle, which is made of a single, unbroken line that loops on itself eternally – 'eternity' in the strictest sense of the word means beginning-less and endless. In Pythagoras's scheme, the second number is what happens once unity becomes more dynamic, once we get 'duality'. This defines the difference between one thing and the other. And that instantly brings about the necessity for the third number which, for Pythagoras, was a most important number. In fact he called it the first 'real' number.

Pythagoras is describing here the traditional view, still alive today in the surviving primary cultures of the world, of how life 'becomes' – how it unfolds from an indivisible unity – its Oneness – into a multiplicity of many, all of which can only be connected by there being a third element of relatedness. In other words, for the one thing to be *known* by another there must be a linking relationship, a ratio or 'joining together'. The Greek word that means 'joining together' is *harmonia*. Just like the ancient Egyptians and their concern for the goddess Ma'at, Pythagoras and much of the ancient Greek world that followed

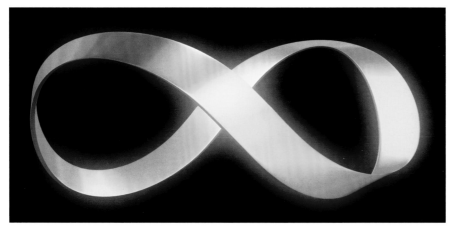

him knew very well that without *harmonia* there is no possibility of relationship between the one and the many and therefore no possibility of unity and wholeness. I am no astrophysicist, but these terms ring with a certain familiarity these days. They could surely all be applied to what has so recently been discovered from quantum physics about the very nature of matter, as I will explore in the next chapter.

Hearing and seeing the grammar of harmony

Pythagoras would no doubt have been pleased had he been able to see what we can see with our electron microscopes, but he explained his insight best with his study of music. Musicians in his own day had known for a long time that if strings of different lengths are plucked together they make pleasing harmonies, but it was supposedly Pythagoras who worked out why. According to a famous Middle Eastern folk tale, Pythagoras was one day walking past a blacksmith's workshop when he heard the sounds of different hammers pounding the anvil. Mostly they just made a noise, but every so often he noticed they fell into a sequence that produced something special. When he went inside he discovered that the hammers were all of different sizes and when he measured them, all but one had a particular mathematical relationship. If these hammers struck the anvil in sequence, the notes they produced had a harmony to them. This was because one turned out to be half the weight of the biggest, another was two-thirds the weight and the next was four-fifths the size of the largest hammer. In this way Pythagoras is thought to have defined the octave and how it relates to the third and the perfect fifth. These are the key musical intervals that, for centuries, dictated the entire grammar of Western tonality.

Today music, like everything else, has been subjected to the influence of

A drop of mercury subjected to a stable soundwave. It is forced to form one of the basic geometric shapes known as the Platonic solids. If a different modulation is used, so the drop changes its shape but it always conforms to the same grammar of harmony.

twentieth-century thinking. Modernism pervades music as much as it does everything else. It is perhaps more than coincidence that just as the twentieth-century ideology of Modernism began to sweep away so much traditional thinking in art and architecture, so the likes of Schoenberg and his Second Viennese School began pioneering the idea in music of abandoning traditional tonal harmony, creating instead an 'atonal' approach to musical structure – that is, a system of notes where there is no controlling primary foundation tone or key.

Many composers followed suit and produced some very interesting and moving pieces of music but inevitably their efforts led to more and more extremes of experimentation, particularly in the 1960s and 1970s. Composers like Stockhausen and the Darmstadt School, for example, produced music that is so unmelodic and so intellectually taxing that it remains completely incomprehensible to the majority of people. Like much of the cutting-edge architecture of the time, it is 'clever'. It tends to appeal to the head and, more often than not, only the cleverest of heads, so it carries with every dissonant turn it takes the implication that we have to be just as clever as its composer to understand it. This is an idea entirely at odds with the root chord of the traditional approach to harmony, which recognizes that we do not 'think' music, we resonate with it and 'feel' it.

Plato's One and the Many

Two thousand five hundred years ago much of Pythagoras's legacy was inherited by another of the great pillars of Western civilization. His work is of such importance that there are still many scholars today who will tell you all subsequent philosophers have offered little more than a set of footnotes to his great learning. This was Plato, who was born around 428 BC, some seventy years after the death of Pythagoras. Plato became a student of Socrates in Athens and immersed himself in the study of timeless truth, his aim being to seek a clear view of the relationship between a particular thing and its existence in the world as a whole. He held that there was an essential relationship between all of the multiplicity of life that buzzes and whirls around us and the unity of the entire universe that supports it. As he put it, 'the ancients who were superior to us and dwelt nearer to the gods have handed down a tradition that all things are said to consist of a One and a Many and contain in themselves the principles of Limit and Unlimitedness'. This is why he claimed that the highest study of all is the study of the harmonies of music and the ratios of geometry because they represent the pattern *within* humanity. He taught that through a strict geometric code there is scope for unlimited variety and many profound expressions of beauty, but only so long as they keep within the limits of the unity of the whole. Without the whole in balance, neither a work of art nor life itself can sustain itself in a durable and healthy fashion. In Plato's scheme, the more out of balance a system becomes, the more radical the compensations, until the entire system cannot hold its centre and it falls apart. Geometry was so important to Plato that he is said to have erected a sign over the entrance to his academy in Athens that warned students not to enter unless they knew geometry.

This was the foundation principle behind what we call the 'classical tradition' and it went on to be the bedrock of learning throughout the high points of Western civilization. Take this observation made by Marcus Aurelius, the first-century Roman emperor who became one of the most important of all the so-called Stoic philosophers: 'All things are linked with each other and bound together with a sacred bond. Scarce is there one thing that is foreign to another. They are all arranged together in their proper places and jointly adorn the same world. There is one orderly, graceful disposition of the whole. There is one God in the whole. There is one substance, one law and one reason common to all intelligent beings and one truth, as there must be one sort of perfection of all beings who are of the same nature and partake of the same rational power.'

This is little more than a pen sketch of the foundations of ancient thought as I have come to understand it, but I wanted to give this little snapshot of how the ancient world saw things from a philosophical point of view because, although it all happened a very long time ago, it is far from 'ancient history'. Fashions may have changed and standards of behaviour and human sensibilities may have become more acute, but the world itself and the way in which life manifests itself in the universe has not changed at all. The ancient deep contemplation of what life reveals itself to be is extraordinarily precise and, in my view, what they considered their experience to be rooted to is as relevant to our experience today as it always has been.

Each culture and tradition characterizes the spiritual dimension of our existence in a slightly different way but, as the Sufis say, although there are many lamps it is all the same light. What I hope is that what we have explored here gives sufficient context to what follows, which is not just the vivid demonstration of how the grammar of harmony and the geometry of the universe fit together, but how our ancestors took this understanding and expressed it so brilliantly in their greatest works of sacred architecture. I want to show the critical importance of harmony to the health of the human condition and the sustainability of the natural world upon which that human condition depends.

The grammar of harmony

In 1957, in Turkey, the explorer, James Melaart, discovered the remains of one of the oldest cities in the world. Çatalhöyük dates from 7500 BC and among the oldest artefacts he found there was a Neolithic series of murals that had lain undiscovered for nearly 10,000 years. These murals contain the self-same patterns that are still woven today into kilims, the small carpets that are made in the region. It is astonishing to watch these rugs being woven in places like Silifke in Turkey, where people are still semi-nomadic. They carry their little looms as they travel and the patterns they weave are not written down anywhere. The women who count the knots simply know them. Language is language, whatever form it takes and so, because a child is born with the natural ability to learn language, the young girls who sit alongside their mothers pick up these patterns as they do the words of their mother's tongue. When asked how they know what pattern to produce these children are unable to say, and yet they can turn from weaving one intricate pattern to the making of another without ever referring to a design somewhere written down. To watch them work is to

Nothing exists for its own sake, but for a harmony greater than itself which includes it. A work of art which accepts this condition and exists upon its terms honours the creation and so becomes a part of it.

WENDELL BERRY

witness timeless wisdom in action that has been transmitted through thousands of years by a tradition that guarantees that what they produce is as vibrant today as it was when it was first conceived.

Something as practical as a Persian carpet shows how these patterns have remained true to what they once represented. The textile museum in Washington DC has a vast collection that spans 5,000 years of history. Many of them are the famous Persian 'magic carpets'. This name has more meaning than might at first appear. They were called 'magic carpets' because they have the capacity to transport us to another place. The carpets represent the designs of Islamic gardens which are based upon the grammar of harmony we are about to explore. So, in effect, these carpets were mobile gardens. When people journeyed through the desert they would take these carpets with them and roll them out under their tented pavilions at night so that they could once more benefit from contact with a garden. Not because they wanted to have flowers in the desert, but because a garden in the Islamic tradition is symbolic of the inner sanctum of the heart. The soul itself is seen as a garden, the garden of paradise, and so the 'magic' carpets transported the desert traveller to humanity's true home, to the paradise within.

I have gained a great deal of insight into the meaning of Islamic symbolism

Shrine of Jalaluddin Rumi, the great thirteenth century Sufi poet, Konya, Turkey, adorned by sumptuous Turkish carpets. Rumi was a supreme poet who created the Sufi brotherhood known for their distinctive spiritual dance, commonly known as whirling Dervishes.

An Afghan girl weaves a kilim carpet in Ghor province. Traditional craftspeople often produce goods of great refinement using very basic tools. The skill resides not in the tool but entirely in the person.

and of the universal geometry that constitutes the grammar of harmony from the renowned world authority on sacred architecture, Professor Keith Critchlow, who helped me found and direct my School of Traditional Arts in London some twenty years ago. An architect himself, he has dedicated his life to the study of the ancient principles and how they work in sacred buildings. He once told me a wonderful old Chinese story about an ancient sage who was asked by his young pupil to draw a picture of the universe. The old man hardly hesitated before picking up his brush and promptly making three swift strokes on the page. First he drew a circle, below it a triangle, and below them both a square.

Keith Critchlow has pointed out that all of the many designs in Persian carpets are elaborations of the interplay between those three basic shapes: the circle, the square and the triangle.

As we already know from Pythagoras, in all sacred traditions the circle is symbolic of the unbroken unity of Heaven. The square is symbolic of the materiality of our earthly existence and the triangle is symbolic the world over of human consciousness.

As that apocryphal Chinese sage knew well, the circle is the best of all starting points because it contains all of the properties that build and support the world. This is a truth. We can learn a great deal about the geometry of the universe simply by drawing a circle. If you were to do so using a pair of compasses and then were to place the point on the circumference of your circle you could draw a second circle that would have exactly the same proportions and overlap the first exactly halfway across. Where they overlap, you would then see a third shape. This is the progress Pythagoras was describing – from the unity of Oneness we get the duality of Two and thus that first 'real' number, the third, relating figure. Pythagoras and Plato both considered this elliptical, rugby-ball shape to be significant because it contains all of the most important geometric properties that make up the grammar of the natural world. The shape itself occurs many times in Nature not just in material form, but in the way energy flows and finds balance in the world. If two bubbles of the same size fuse together they will do so in just this way, with the one sphere overlapping the other at precisely the point where together they produce this same special shape in the middle.

The shape is very familiar to us all. It has been given many names throughout history and it has been used as a central symbol in many cultures around the world. It usually suggests the same thing. Pythagoras thought it looked like the 'measure of the fish'. Perhaps this was an influence on the early Christian Fathers, who thought it looked like the bladder of a fish, which is in Latin 'vesica pisces'. The Romans themselves thought it looked like an almond, and a common name for it today is the Italian word for almond, 'Mandorla'. But its shape has been found in much earlier civilizations. In very ancient traditions the shape was associated with the goddess Venus, symbolizing the female organ of birth – the doorway or window between two worlds. In ancient Egypt it was turned on its side to form the great eye of Horus. In the Judaic tradition it has always been used to describe the shape of Noah's Ark – the Ark of the covenant being the very symbol of the wisdom of the world. The vesica is found woven into the fabric of Buddhist architecture and Christian symbols. Christ is very often depicted within a vesica; the floor plans of many churches and cathedrals

Christ in Majesty depicted within a vesica above the Royal Gate of Chartres Cathedral, France, flanked by the four Evangelists. The ratio between the width and height is the square root of 3 or 265/153. 153 figures in John's Gospel was the number of fish caught miraculously when the risen Christ appeared before the Disciples.

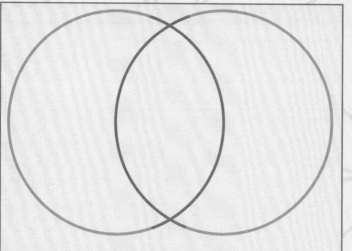 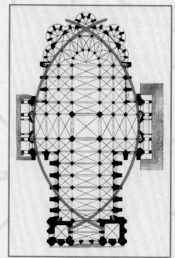 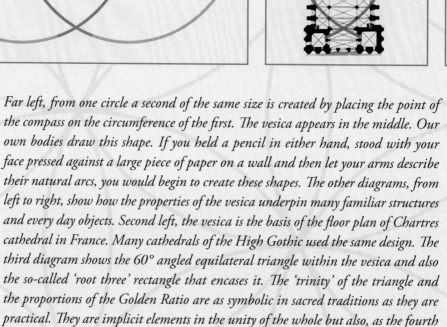

Far left, from one circle a second of the same size is created by placing the point of the compass on the circumference of the first. The vesica appears in the middle. Our own bodies draw this shape. If you held a pencil in either hand, stood with your face pressed against a large piece of paper on a wall and then let your arms describe their natural arcs, you would begin to create these shapes. The other diagrams, from left to right, show how the properties of the vesica underpin many familiar structures and every day objects. Second left, the vesica is the basis of the floor plan of Chartres cathedral in France. Many cathedrals of the High Gothic used the same design. The third diagram shows the 60° angled equilateral triangle within the vesica and also the so-called 'root three' rectangle that encases it. The 'trinity' of the triangle and the proportions of the Golden Ratio are as symbolic in sacred traditions as they are practical. They are implicit elements in the unity of the whole but also, as the fourth diagram shows in the cross section of Chartres, the equilateral triangle also suports

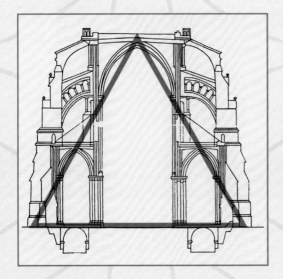

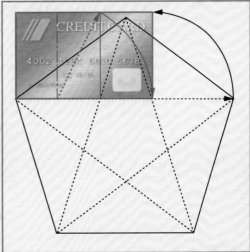

the entire vertical structure of the building. The diagrams on these two pages sit on top of a background made up of the pattern created when six circles surround the first. These 'six days of creation', as the Bible calls them, create 'the flower of life', a familiar tile pattern found in ancient Greek architecture, Islamic design and on the floor of many Christian churches. The flower of life contains in turn the coordinates of the five-pointed star, far left. These familiar shapes and patterns make up the grammar of harmony and have been used to attract the eye for centuries. Even the humble credit card employs the ratio of the Golden Rectangle, seen here in the five-pointed star. The lines of this geometry remain invisible in the patterns of traditional art. In the view of traditional philosophy, in Plato, for instance, symbolically they were used to represent the underlying structure of reality upon which the cosmos materializes. The word 'cosmos', incidentally, means 'adornment'.

 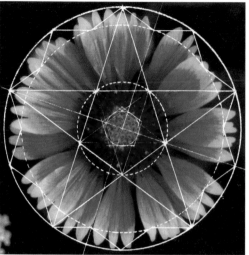 

The grammar of harmony at work. From 'the flower of life' comes the pentagon and the five-pointed star, far left. As we shall see, Venus describes these shapes in the skies above us every eight years (or thirteen Venusian years). I wonder how much of a coincidence it is that the self-same five-pointed star and the relationships we will soon discover between Earth's orbit and size and that of her nearest neighbours are to be found in so many plants and flowers on the ground around us? In the image third from left, the two opposite spirals of the Fibonacci sequence are clearly visible in the head of a daisy.

are laid out within its structure and the light that streams in through so many church windows is framed by it too. It is even there on many a modern car bumper – the two lines that form the fish emblem on a sticker used by many Christians who wish to declare their faith, but who perhaps are unaware of the geometric reason why that shape was chosen.

If you were to draw a line across the middle of this almond shape and then draw two more down from the top so that each one meets the baseline as it intersects the outer walls, you create a perfect equilateral triangle. This is a vitally important shape in geometry. Not least because it is one of the strongest, load-bearing shapes in all architecture. Plato called it the most beautiful of all triangles.

Constructing the equilateral triangle allows the construction of a square and from the square and the triangle comes a special rectangle which, down the ages, has also been profoundly symbolic. It has long been known as the Golden Rectangle. Technically it is called a 'Root Three Rectangle' and it is special because the ratio between the two lengths of the sides is 1:1.618. This may not look a very remarkable set of numbers, but this single ratio is a very significant relationship indeed.

In the twentieth century it was given the name *Phi* by an American mathematician, Mark Barr. *Phi* is the first Greek letter in the name Phidias, a sculptor whose work stood in the Parthenon above Athens and, like the Parthenon itself, the beauty and balance of Phidias's sculptures depend very much upon the use of this ratio of 1:1.618. The Greeks themselves referred to it as the 'Golden Ratio' or 'Golden Mean' and it has become famous of late because of popular books and films like *The Da Vinci Code*. Even so, it has long been understood

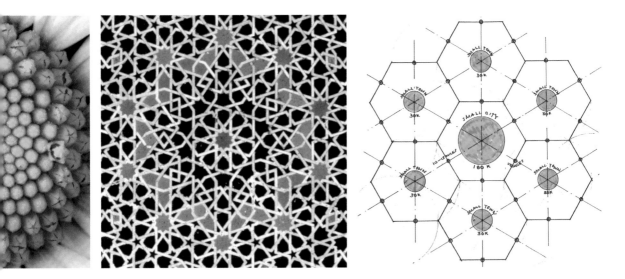

and employed in geometry and architecture because it is the ratio that rather miraculously governs the way that organisms unfold. Even by looking at something as small and as common as the head of a daisy, this ratio can be seen at work. The seeds swirl from the centre in a far from random vortex. The lines travel in two directions that have a precise mathematical relationship. The number of seeds we see swirling in one direction are related proportionately to those travelling in the other and that proportion is the Golden Ratio.

The sequence of numbers that describe this proportionate relationship has, for a very long time, been known as the Fibonacci sequence, named after the thirteenth-century Italian mathematician who made a long study of the way the number of rabbits increases in every generation. He noticed that the way rabbits multiplied followed the same sequence that plants conform to when they sprout new leaves or when a tree produces new branches. The sequence starts with one pair, then branches to make two pairs. Then, as the gestation periods of the different pairs progress at a different pace, so the branching follows a curious multiplication from 2 to 3 to 5 to 8 to 13. This sequence is more related than at first appears. Each is the product of adding the preceding two. What is even less obvious is that if any of these numbers is divided by the one that precedes it, the result hovers around the same number, the famous 1.618. The bigger the numbers become, the closer their division gets to this golden number, a number that Johannes Kepler called a 'precious jewel.'

There is an elegance to the Fibonacci sequence. If each of the numbers is measured either in inches or centimetres and plotted out on a piece of paper it produces a pattern of boxes. Joining the corners of those boxes with a single continuous line produces a very familiar shape indeed. Not only the spiral of

Such natural patterns have always been used in the Islamic traditional crafts, second from right, to depict the relationship between the order in Nature and the organic process of unfolding. It is this same geometric relationship that dictates a very modern application, far right. This is a sketch from my Foundation for the Built Environment demonstrating the idealized pattern underlying what we have called 'walkable towns' where buildings are clustered in spiral-like arrangements around the intersection of roads or pathways, creating a series of village centres. This is the pattern behind my development at Poundbury.

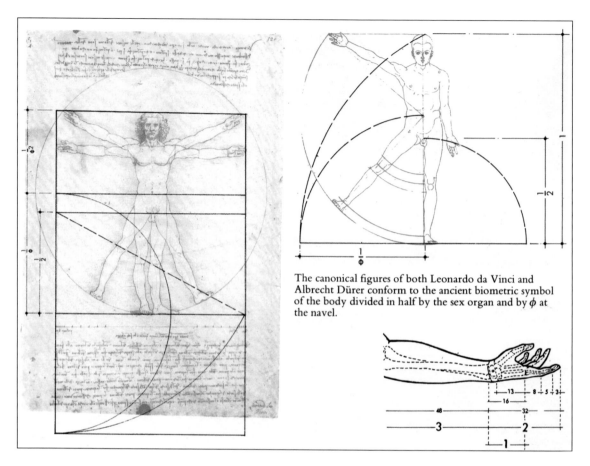

The canonical figures of both Leonardo da Vinci and Albrecht Dürer conform to the ancient biometric symbol of the body divided in half by the sex organ and by ϕ at the navel.

The geometry of flowers is the geometry that controls the growth and proportions of our own bodies. As Da Vinci demonstrated, these ratios are all related to the way the human body describes the circle of a given perimeter and the square of the same length. Many cultures have seen the squaring of the circle as the meeting of Heaven and Earth, where the ideal and the actual, spirit and matter come together.

the seeds on the daisy head, but the shape found all over the natural world, the sort inscribed on the shell of a snail or, indeed, the shape that our forefinger makes when a human hand is clenched in a fist. The same numbers are always at play. Each section of the human finger, from the tip to the wrist, is proportionate to the next section, according to the Fibonacci sequence, just as the proportions of the rest of the body are too – from the nose to the neck, from the neck to the chest, and so on. Even as we grow, these numbers play their part. The way our teeth grow follows the general pattern 1, 2, 3, 5, 8, 13.

This spiral shape is also present in every river of the world and for very good reasons. I was fascinated when I first came across the work of the Austrian forester Viktor Schauberger, who demonstrated in the 1920s that rivers do not flow as a block of water but via spiralling vortices. Our blood supply does the same. In this way the friction that the blood would cause as it moves through our bodies is reduced and, indeed, the immense pressure on the veins and arteries. If the blood didn't do so, our veins would burst and our fingertips would burn to a frazzle.

Schauberger also demonstrated that for a water vortex to spiral as it does, a similar movement of water or air must be on the move in the opposite direction. Incredibly, this means that, in effect, a river flows in both directions, creating a circle from source to source rather than a straight line down to the sea. Evidence of this came some years ago in Leipzig after a great storm, when a bridge with all its historic decorations was swept away in floods. When the water and storms subsided, much to the amazement of the engineers who dredged the river, the pillars of the bridge were found not downstream but a mile above where the bridge had stood.

Schauberger's work was misunderstood for much of his life. After the Second World War he was taken to the United States, but quickly returned to Austria because his theories were thought not to make sense. Only now are they being put to practical use, demonstrating just how brilliant his observations were.

Compare this image of a whirlpool with a satellite image of a hurricane over Florida and you will see the same pattern. Note the large fish in the centre.

They are helping river engineers, for instance, to improve flood defences without damaging the health of rivers, which is something that has not been easy to achieve using conventional methods and, as I will explain later, when we come to the chapter on solutions, others are now taking this technology and producing some very exciting devices that use hardly any energy to move colossal quantities of water. There is also research into developing propulsion systems for aircraft and for boats. We should look closely at what this pioneer Schauberger had to say because there may be many more useful implications for the way we develop the technology of tomorrow. We should certainly take note of the key principle of Nature's approach to using energy, which he never tired of pointing out. Unlike human technology, which is dependent upon combustion to produce energy, Nature draws energy *into* herself – sunshine into leaves, nutrition from the soil into plants, and so on. Even the way fish move through water corresponds to this opposite way of using energy. Schauberger showed that the swiftness of a fish and the ease of this action depends upon yet more spiralling vortices of water, this time tiny ones, created in the wake of each scale on the fish's body. These are what make it possible for a fish to hold its position with little effort, even against the force of a fast-flowing stream. Schauberger put it rather beautifully: 'The fish does not swim in the river, the river swims the fish.'

Whether it be the shape of a plant, the arrangements of petals in a flower, or the harmonies of the music made when it is constructed according to this numerical relationship, we call it 'beautiful'.

Whole books have been written on the magic of the Fibonacci sequence, but I do want to draw attention here to the fact that we respond to the patterns it creates in a certain way. The same system of numbers that describes the spiralling vortex of water is also found in the way music works. I don't want to get into the technicalities, but simply notice that those same numbers – 1, 2, 3, 5, 8, 13 – are the ones that describe what is known as the chromatic scale in

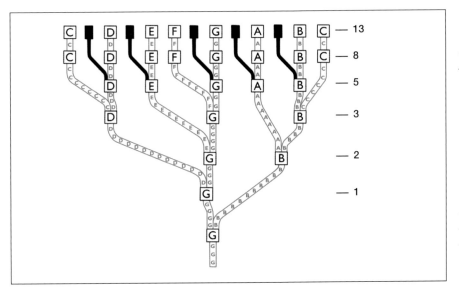

music – that is, the combination of black and white notes on a piano keyboard that play the octave. There are eight notes in the octave, which includes a harmonious third note and the so-called perfect fifth. If you play all of the notes, black and white from, say, one G up to the next G, then you play the thirteen notes of the chromatic scale. And if you work within the Indian system, then there is another scale above the chromatic which is known as the Śruti and that involves the next number after 13 in the Fibonacci series, 22. There are many other examples. The golden relationship has even been found to occur between Quarks and subatomic particles. The theoretical physicist and engineer, Professor M.S. El Naschie, has conducted some fascinating studies in this complex field.

On the mathematical level this is all very interesting, perhaps it is even entertaining, but my point is that, wherever it appears, we find this shaping and patterning so naturally pleasing. Whether it be the shape of a plant, the arrangement of petals in a flower or the harmonies of the music made when it is constructed according to this numerical relationship, we call it 'beautiful'; we have a sense of an immeasurable quality; we feel something special and indefinable when these numbers are at play. There are even well-documented experiments where different groups of people have been shown pictures of different faces and asked individually to say which ones they find more beautiful and generally they tend to settle on those images where a person's features conform most readily to the ratio of the Golden Section – the relationship, for example, between the width of the eyes and the length of the nose, or the width of the mouth to the width and height of the forehead; even the spacing of the

RIGHT: *Newgrange,*
built in County
Meath, Ireland,
around 3200BC.
The entrance stands
behind one of the 97
kerbstones, decorated
with the familiar tri-
spiral design. The
symbol is commonly
thought of as Celtic but
this stone predates the
Celts by a thousand
years. Sculpted stones of
the same period found
elsewhere in the UK
demonstrate that the
generations who built
Newgrange understood
the importance of what
we call today the
Platonic Solids, the five
essential geometric
shapes – the cube,
the pyramid (or
tetrahedron) through
to the twenty-sided
icosahedron. These are
the foundation shapes
of all matter.

teeth. It seems that we resonate naturally with the proportions that reflect this golden proportion. No wonder the banks decided that credit cards should be golden mean rectangles or that Apple decided to use this ratio to create the shape of its first iPod.

From such evidence I cannot help concluding that the idea of beauty as 'being in the eye of the beholder' is perhaps the other way around. Perhaps beauty itself speaks rather than something made by us, nor is it something we have to think about. It is something we sense without thinking, being there already as an element of the innate 'patterning' of consciousness. Certainly it was the ancients' view that when we recognize beauty we come into direct contact with the archetypal patterns of creation, and the more we contemplate them, the more sensitized we become to the imprint of the very ground of our being. So it is vital that we *do* come into contact with them, or we suffer a loss that is very detrimental to our well-being.

Stradivari and the music of the spheres

This connection with the structure and behaviour of the universe is to be found in nearly all of the traditional crafts, and sometimes unexpectedly so. For example, I was captivated when I came across the work of a young geometer called John Martineau while he was studying at my School of Traditional Arts some years ago. He decided to make a close study of how the orbits of the planets relate to each other and how the patterns that can be drawn from them fit so precisely with things made down here on Earth. He found many rather beautiful relationships but, as an example, consider the one that exists between the orbits of Venus, Mercury and the Earth.

The five-pointed star

Ancient astronomers long ago discovered that Venus makes an intriguing journey across the sky. It takes eight years – or thirteen 'Venusian years' (note the relationship) – for it to return to exactly the same position it is found in today, and I mean exactly. It is such an accurate cycle that many cultures down the ages and in very different parts of the world have all noticed this and used Venus as a way of regulating their calendars. For example, there is a vast prehistoric site at Newgrange in Northern Ireland which, for hundreds of years, was thought to be some sort of prehistoric burial mound. More recently studies have shown that, whatever else it might have been used for, Newgrange is also a very precisely aligned 'Venus trap'. Once every eight years, and only for a very

short period on that one day, the light cast by Venus when it appears at sunrise as the Morning Star passes down the long entrance tunnel and hits the wall at the back of the inner chamber that lies at the very centre of the mound. It does this so regularly that the researchers who carried out their study claim its accuracy was only slightly improved with the invention of the atomicclock.

As ancient astronomers charted Venus's progress through its eight-year cycle, they discovered that it describes a swirling rose-like pattern. The illustration was made by John Martineau as he tried to verify how ancient cultures devised the symbols that are still so familiar today. The Earth is at the very centre of the picture. There are moments when the line comes closer to Earth and then moves away again, creating a circle of five petal-like shapes. If we were to join the tips of that pattern together, as ancient astronomers clearly did, then what is revealed is a shape familiar the world over, the endless line that forms the five-pointed star. It is a shape that contains some breathtaking secrets.

The orbits of the planets are not perfectly circular, but it is possible to refine their elliptical shapes without altering their length so that they become perfect circles. Such a circle is called the 'mean orbit' of the planet. John Martineau found that putting scaled drawings of the mean orbits of the Earth and Mercury together on a piece of paper reveals an extraordinary correspondence between them. In the image (above right) the Earth's orbit is the bigger circle that contains the five-pointed star. The smaller circle is the mean orbit of Mercury, which sits within the orbit of the Earth in such a proportion that it fits exactly over the pentagon at the heart of the five-pointed star. If that were not itself astonishing enough, the same thing happens if a scaled drawing of the actual physical body of the Earth is overlaid with a scaled image of the actual physical body of Mercury. Mercury, once again, sits inside the circle of the Earth's circumference in exactly the same proportion. The pentagon shape at the heart of the five-pointed star is once again enclosed by Mercury's circumference.

This may, of course, all be a coincidence, but such is their precision it does begin to challenge the popular notion that we live in an accidental

Although the Earth and all other planets circle the sun, from our point of view the planets appear to dance across the fixed zodiac of the sky. The word 'planet' means wanderer. This is the dance of Venus as seen from Earth, charted over its eight year cycle creating the heart-shaped set of five petals from which so many familiar geometric shapes are derived.

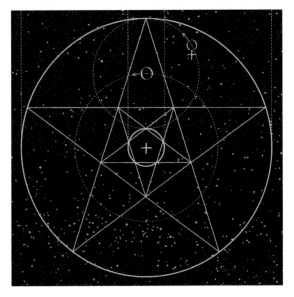

universe, especially when the same things happen at levels of the material world that we cannot see without the aid of a microscope. For example, we are now all familiar with the double-helix shape of the DNA molecule. Deoxyribonucleic acid is present in nearly all forms of life and transmits all of our genetic information from one generation to the next. The less familiar image of the molecule is the view taken from the top of the double helix. When we look down on the molecule through a microscope the image is not dissimilar to the one of Venus's journeys across the night sky. It is a swirl of patterns with ten protruding petals. If every other petal is connected by a series of straight lines, once again what emerges is the same five-pointed star. This five-pointed star, found in so many petal arrangements on flowers, appears constantly in the patterns and architectural designs in Islamic buildings as it does in Christian structures and it also underpins the structure of some of the most familiar man-made objects we know.

Take the image of a Stradivarius violin, for example, and place it in a circle that already contains a five-pointed star and the impact is just as breathtaking. All of the key proportions of the violin fit the geometry of the star perfectly. And notice that the base of the violin is also the product of those two overlapping circles that create the mandorla shape with which we began this demonstration. For the ancients the two overlapping circles also represented the Sun and the Moon. The Sun, of course, is much bigger than the Moon, although from Earth this does not always appear so. Every so often we are still drawn to marvel at a total eclipse of the Sun when the Moon, seen from Earth, is exactly the same size as the Sun. So, even in the sky, the grammar of harmony

ABOVE LEFT: *The relative mean orbits of Mercury and Earth superimposed over each other. The Earth's orbit contains a five-pointed star and the circle of Mercury's fits exactly over the inner pentagon of the star.*

ABOVE: *A less familiar view of the DNA molecule from above reveals the ten points on its outer rim which allow two five-pointed stars to be drawn within it.*

Some of the most familiar objects depend upon the geometry of the universe. Here the structure of a Stradivarius violin fits perfectly within the grammar of harmony. Even something as familiar as the front door on an English Georgian house, like the one on Number 10 Downing Street, accords with the interplay between circles and equilateral triangles.

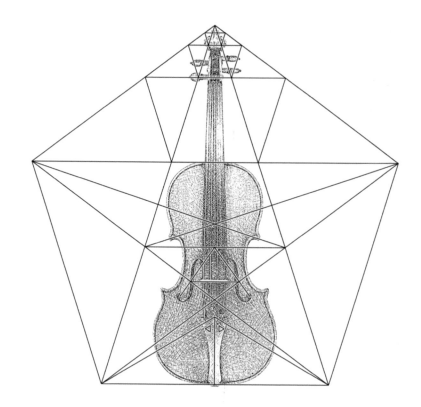

is at play. This is all pretty remarkable evidence that there is a mysterious unity about the patterns found throughout the whole of creation. From the smallest of molecules to the biggest of the planetary 'particles' revolving around the Sun, everything depends for its stability upon an incredibly simple, very elegant geometric patterning – the grammar of harmony.

The weave towards modernity

This geometric code that I have called the grammar of harmony was evidently understood by every one of the major civilizations of the world. The temples of India reflect it profoundly. Many of them follow a similar design. At the centre sits a dark chamber and this is surrounded by a series of rooms that become lighter as they get nearer to the outside world. The symbolism is missed by most, but the point here is that all of creation bursts out of what the mystics of India call the 'uncreated light' of the central unity. From this unity flows all of the teeming multiplicity of existence, symbolized by the rich decoration and intricately carved ornamentation of the temple's outer walls. Again, such temples are models of the universe, both its outer aspect and its inner one.

The same application of geometry took on different forms in ancient China, but the purpose of the message remained much the same. Consider this statement that could so easily be made by a quantum physicist of today, but in fact was composed in China over 2,000 years ago by a near contemporary of Plato, the Chinese sage Lao Tzu, in one of his lesser-known collections of observations, the *Hua Hu Ching*: 'The world and its particles are not separate, isolated things but rather, one small particle contains the nature of the world just as the world contains the nature of each small particle; the nature of each is the same. The apparently single event is but a variation and segment of the great whole and the great whole is the combination of all single events. Thus, the single events contain the life experience of the whole.'

Salvaging the treasures

This is the perspective that ruled the outlook of the ancient world. What I now want to explore briefly is what happened to this way of going about things when the classical world finally collapsed. The deep understanding of the meaning of harmony did not die out. It became the rich and fertile foundation of two great pinnacles in the history of civilization in the East and the West. Tracing

Shri Swaminaray Mandir Temple in London. A modern building which reflects the geometry of the ancient temples in India and Bangladesh. The basic plan of any Hindu temple is a Mandala, an Eastern sacred shape that combines the circle and the square, symbolizing both the outer universe of the macrocosm and the inner space of the the spiritual microcosm.

the journey it made from the collapse of the classical era to our own time will hopefully demonstrate just what it was that our culture lost when the prevailing world view shifted so significantly at that linchpin period in the seventeenth century when the modern era was born.

It was the Arab world that salvaged much of the treasure from the ancient world. Slowly it infused Arab thinking so that when the great Abbasid Empire rose to prominence from the eighth century onwards the principles of harmony, balance and unity were central to the vast outpouring of craftsmanship and scholarship that characterized what has become known as 'the Golden Age of Islam'.

True, without falsehood, certain and most true, that which is above is as that which is below, and that which is below is as that which is above, for the performance of the miracles of the One Thing. And as all things are One, so all things have their birth from this One Thing by adaptation. Its power is integrating, if it be turned into Earth.

The Emerald Tablet of Hermes

The vast Muslim Empire that created this golden age eventually controlled territory from Northern India up to Samarkand and from Iran and Turkey across to the Iberian peninsula of Spain. Its epicentre was Baghdad, which enjoyed a spectacular flowering in scholarship and an approach to art and design that fused Arab thought and invention with that of Persia, Egypt, Europe and the Far East. Four hundred years before Europeans discovered the secret, the Abbasid culture had mastered the Chinese art of making paper and this proved as revolutionary an innovation as the internet has been to us. Whereas in

Northern Europe the written word had still to be transmitted on expensive skins of velum, books of paper were soon conveying learning to all corners of the Muslim world. It is estimated that within 200 years of the death of the Prophet Muhammad, books were available in libraries that peppered every major city of the growing empire. By the middle of the thirteenth century there were thirty-six libraries in Baghdad alone, where it was possible to read books

The Malwiya minaret of the Great Mosque at Samarra in Iraq, completed in 851AD when Samarra was the capital of the vast Abbasid Empire. Its unique spiral ascending walkway is called the 'Malwiya' which means 'snail-shell'.

One of the earliest pictures of an astrologer using an astrolabe to calculate the position of the stars. In this thirteenth century Arabic manuscript he is upper right, casting the horoscope of the child born in the centre panel. Astrolabes were developed in antiquity and used widely in the Muslim Empire for navigation and for finding the direction of Mecca. From spherical astrolabes came the modern clock.

on history and poetry, Greek and Islamic philosophy, mathematics, astronomy, and medicine.

Córdoba, in Southern Spain, was the magnificent capital of the Western wing of the Abbasid Empire and it became the gateway through which so much of this treasure passed on its journey to inform the West. Europe at the time struggled in relative darkness. The very biggest library in Christendom at the start of the eleventh century could only boast around 1,000 books, whereas the great library at Córdoba alone is said to have contained 400,000 volumes. Such immense libraries attracted scholars and they came from far and wide to Córdoba, bringing with them many ancient Greek texts that had been salvaged from the ruins of antiquity to be translated into Arabic and Hebrew. These scholars advanced their knowledge in science, natural history, geography, law, history and medicine. Their studies covered everything from agriculture to building design and they made tremendous advances in optics and engineering. Their mathematicians adopted what they called 'Indian numbers' – what we know today as our own 'Arabic' system – which offered a much faster way of adding up and subtracting than the cumbersome system offered by Roman numerals. They invented algebra. Using a Persian device called an astrolabe, the scholars of Córdoba charted the skies, studied astronomical phenomena like eclipses and comets and calculated the circumference of the Earth to within a few thousand feet. Their studies in medicine created standard texts on the subject and they invented hospitals equipped with what amounted to accident and emergency wards. Their business acumen led to the development of sophisticated new business practices that are now common to the world – the notion of partnerships, the use of credit and the idea of banks exchanging currency. And, to make life a little more pleasant, they invented mint-flavoured toothpaste and sweet-smelling deodorants, and even mixed sweet syrups with drugs to help the medicine go down. It is, however, the way they integrated into their culture the patterning of Nature that concerns us here.

The patterns of architecture and decoration that so defined this medieval culture depend upon the so-called 'Seven Sacred Principles' of Islamic architecture, the chief one being *Tawid*, or unity. Five times a day the entire world of Islam turns to face Mecca where, in the centre of the central mosque, stands the immovable cube of stone, the Kab'ah. The emphasis in this act is unity. It is the single testimony upon which Islam rests: that there is no god but God, who is the God of all, or Allah. In committing to this single testimony, all Muslims are unified and this unity of all things is expressed very visually in Islamic architecture. The intention is to make it perceptible at all levels of the built environment. On every wall of every room, in every building and in whole cities, the aim was to create a sense of wholeness, the unity that rests in the heart of every man and woman.

A young boy's vision

This learning inevitably began to travel beyond the bounds of the Muslim world. Many Christian scholars made it their business to get as close to this culture as they could. Sometime around 940A.D. one of the most remarkable

Decoration above the Western Door, the Great Mosque at Córdoba. The Islamic decoration and patterning survived the demise of Muslim rule in the thirteenth century when the King of Spain converted the mosque into the present day cathedral.

of them was born in the town of Aurillac, nowadays the umbrella capital of France. His name was Gerbert and being a bright boy of a poor family from the Auvergne he entered a strict Benedictine monastery and slowly rose through the ranks of the church. It is said that Gerbert visited Barcelona as a young man, benefitting from one of the best-equipped libraries in the Christian world, at Vic in Catalonia, and maybe he even went to Córdoba itself.

Certainly by the time he returned to France and headed North, first to Rheims and finally on to Rome, where he eventually became the first French Pope, Sylvester II, Gerbert was well versed in Plato's geometry, the mathematics of Euclid and the dexterity of Arabic algebra. He brought with him strange instruments, like the Chinese abacus and the Persian astrolabe, and with them came the Indian concept of the Zero, a new understanding of prime numbers and coordinate equations and an insight into the study of proportion, all of which enabled musicians to start tuning instruments more precisely and to play with both harmony and dissonance in a way that they had never done before.

Perhaps if he did make it to Córdoba the young Gerbert may have walked into the new jewel of the city, the exquisite Great Mosque, which had recently been completed after 200 years' work. By then he would have benefitted from the Islamic study of botany, which taught how everything in Nature unfolds from a single point of origin and that these 'unfoldings' happen according to archetypal, original plans. If he did walk into that magnificent building – one of the 1,000 mosques in the city at the time – he would have perhaps discerned, beyond the rich decoration in jasper, onyx and marble, the way the structure 'interiorizes' a sense of harmony and conveys the subtle and abstract 'structure' of consciousness. We can only imagine the impact its decorations must have had on his eye, all precisely according with a geometry that stresses in such a subtle way the unity of all things. Perhaps such an experience would have stayed with him for life and explains why on his return to France Gerbert sought to inspire the building of some of Europe's greatest treasures.

Building the cathedrals

Professor Keith Critchlow who, as I have mentioned, eventually helped me set up my School of Traditional Arts in London, has made an extensive forty-year study of a cathedral inspired by Gerbert. It was begun by one of Gerbert's pupils, Bishop Fulbert, who had founded the school at Chartres, some 70 miles south-west of Paris, and it became what Professor Critchlow has called 'The Queen of All Cathedrals'. Like all great cathedrals, temples, mosques and holy places, it acts as a bridge between the world of Nature, human society and the domain of the spirit. Keith Critchlow has shown how this cathedral is built according to the same timeless principles of geometry I have just explained, but is completely Christian in the way it embodies the extraordinary fulcrum between the Judaic faith and law of the Old Testament and the 'good news' brought by Jesus Christ. He told me once that more stone was quarried and thrust up into the skies across Northern Europe in the 200 years of cathedral building than was used in Egypt in the 2,000 years it took to construct the pyramids.

Chartres was begun at the start of the eleventh century. It is dedicated to the Virgin Mary, and Professor Critchlow suggests that the symbolism of the entire cathedral is intimately tied up with the medieval development of the cult of Mary – perhaps a renaissance, he claims, of the cult of the Greek goddess of wisdom, Sophia. I can only give a snapshot here of the remarkable scholarship he has achieved, but it is so important that the wider world sees some of what he has uncovered because it demonstrates in physical form the outlook that prevailed in the mood of thought in the Western world 1,000 years ago.

Once again the geometry of the entire building is derived from a circle. Its floor plan is contained within the proportions of a vesica. As the illustration

The Belle Verrière Window in Chartres Cathedral dates from 1137AD. The quality of the staining, particularly the blue, has mesmerized generations of glass makers. It is still not fully understood how the medieval glass makers created such a depth of colour.

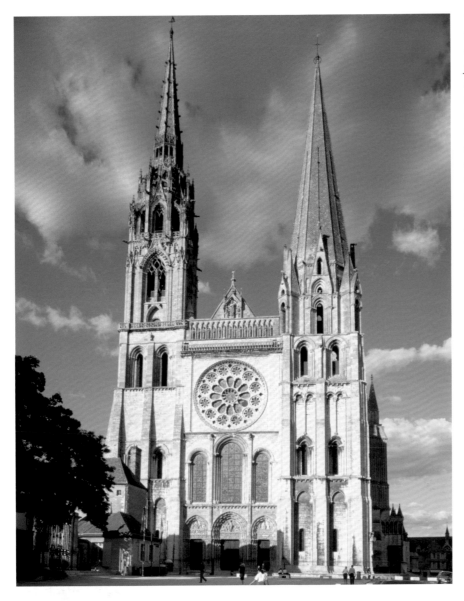

on page 106 demonstrates, the centre point of the vesica sits at the very centre of the building so that the North and South doors are exactly positioned. The windows also conform to this shape. The great Belle Verrière window, for example, which depicts the Madonna and Child, sits perfectly within a vesica and thus perfectly within the floor plan of the cathedral, with every significant point in the design of the window corresponding to key positions in the geometry of the rest of the building. Christ's head sits over the Madonna's heart. As Professor Critchlow has shown, the infant Christ's throat, from which the entire Christian tradition was eventually spoken, falls at the very centre of the vesica and therefore at the very heart of the building. The blue jewel in the Madonna's crown falls on the *rond point* of the cathedral at the end of the choir, which happens to sit immediately below a large weathervane up on the roof in the shape of Gabriel, the archangel who brought the news to Mary of Christ's incarnation. The eight stars that circle the Madonna's head, fall precisely on the eight pillars that surround the altar, and her feet rest on the columns of the doorway of the West entrance. And so the precision of this comprehensive geometry goes on.

The entrance into the building is through the West front, which comprises two soaring towers, one with the symbol of the Moon upon it and one, a significant number of feet taller, bearing the symbol of the Sun. The height of this spire matches the length of the cathedral: 365 feet. The Moon spire is 28 feet shorter, a number linked very closely to the lunar calendar. And beneath them sits one of the most spectacular of all rose windows, symbolizing the uniting of the apparent duality represented by the symbols of the Sun and the Moon. This unifying process is even built into the way the pilgrim was expected to journey around the cathedral. They would enter the building beneath the Moon, passing from the world of time into the timeless, and then progress along the left wall, reading the story of Christianity in the windows of the North side of the cathedral. There are in fact three great rose windows in the body of the cathedral and they were also intended to be read in sequence, along with all of the other layers of symbolism built into the fabric of the architecture.

For instance, there are two gateposts on the building, one on the North door and one on the South. The Northern gatepost carries a statue of St Anne, who, being the mother of Mary, is traditionally the figure in whom the Old Testament ends. After progressing through the building, following the course of Chartres's famous labyrinth with its central chamber containing twelve petals on the floor (exactly at the point, incidentally, where Christ's feet appear in the Belle Verrière window), the pilgrim would pass the Southern door, above which is the image of the fully grown Christ, enclosed within a vesica. On the gatepost of that door Christ is depicted again, carrying in his hands the book of his message to the world, his New Testament. Even here the architecture is full of symbolism. Professor Critchlow points out that the shape of the book is a Golden Rectangle and that it is angled in such a way that one corner rests on Christ's 'navel centre', as it would be described in the Indian, Vedantic tradition, while the upper corner rests on his 'heart centre'. Keith Critchlow has also calculated that this book is tilted at approximately 24½°, which is both the angle at which our hearts lie within our bodies and the tilt of the Earth in relation to the solar axis.

The image of St Anne, the mother of Mary, on one of only two door posts at Chartres.

Quite clearly not an inch of this entire building is left to chance. Every angle and position conveys symbolic meaning. The medieval Christian architects who designed such a breathtaking structure were following the teachings of the mystics of their age and created what seems to me to amount to a profound prayer to all of creation. They made a building that offers us the direct experience of what the ancients held to be our true relationship with the world. To walk around within its soaring pillars and to bask in the gentle light that pours through its exquisite windows is to experience a sense of participation in the very 'patterning' of the soul. No sense here of being a disconnected observer in a dead and mechanical universe.

I have wanted to pay such attention to the principles of the world's sacred geometry because they stand in such contrast to the predominant way in which we view the world today. I am sure many people will say that you cannot organise 'modern' life around ancient, irrelevant concepts, but the point is that the sheer elegance of a building like Chartres Cathedral and the precision of its geometry was only possible because of the rediscovery of classical knowledge which was born of a tradition of wisdom which is not time-specific and

The Carpet Garden at Highgrove, my home in Gloucestershire, England. It was inspired by designs of Turkish carpets which were derived from the patterns of Islamic gardens. I planted fig, pomegranate and olive trees in the garden because of their mention in the Qur'an.

La Primavera, Spring, by Botticelli. This is the complement to Botticelli's other great painting, the Birth of Venus. *They hung opposite each other in Lorenzo de Medici's quarters, depicting the Heavenly and the Earthly aspects of love.*

'historic'. It is timeless and extremely relevant to the way the natural world works today – as it has always been. Nature has not changed her attitude because of fashion.

As our history books tell us, the renaissance in this learning reached something of a zenith a couple of centuries later in fifteenth-century Florence, where figures like the long-forgotten but tremendously important Marsilio Ficino began to incorporate Plato's principles into the established Christian tradition.

Ficino taught the young Lorenzo de Medici in Florence and eventually headed the Florentine academy established by Lorenzo's grandfather, Cosimo. This was a copy of Plato's academy and one of its aims was to translate all of Plato and many other classical philosophers and writers. It was in this capacity that Ficino published one of the most influential books of the entire Renaissance. In short, this book characterized the blend Ficino had achieved that is called today Christian Platonism. Eventually this vision of the world would come up against the growing surge of humanism in the seventeenth century that marks the great turning point in our story here. It was humanism that forced the famous split between Church and State in the seventeenth century and opened the door to those notions that became so prominent in the European Enlightenment. By then, as we see in the next chapter, human nature came to be seen more and more as self-interested. Nature was also transformed into nothing more than the battleground of competing species. This is a very different view of humanity and the natural world from that held by Ficino, who from his base in Florence persuaded many painters, writers and musicians throughout Europe to make a fresh connection with Nature and the eternal principles that she displays.

The renaissance in this learning reached something of a zenith a couple of centuries later in fifteenth-century Florence, where figures like the long-forgotten but tremendously important Marsilio Ficino began to incorporate Plato's principles into the established Christian tradition.

One member of this group was Ficino's friend Sandro Botticelli. He, too, was employed by the Medici and was heavily influenced by Ficino's brilliant mind. It was under Lorenzo's patronage that Botticelli painted two of the world's most famous paintings and both the *Birth of Venus* and *Primavera* clearly display Ficino's call for us to see how intimately connected we are to Nature's harmonic balance. *Primavera,* in particular, could well be described as a painting of the unity in all things. Clever scholarship has revealed just how perfectly the underlying design of the painting depends upon the same geometric proportions and shapes we have already explored – the Golden Ratio is found all over it, for instance. It is also full of symbolic references to Ficino's teaching of Platonic philosophy. And, of course, it depicts the arrival of Spring, the rebirth of natural wisdom. That it was painted for a 14-year-old boy says

Dr John Dee 1527–1608, the great Elizabethan polymath whose library was said to be the greatest in all Europe. James Burbage consulted Dee on the matter of acoustics before he constructed his new theatre in London, the Globe, where Shakespeare's plays were performed.

something about the education the young Lorenzo received at the knee of the great Ficino.

Ficino's highly influential book *Platonic Theology* found its way to England shortly after it was published and eventually, some ninety years later, into one of the most important libraries in Elizabethan London, owned by Dr John Dee, Queen Elizabeth I's astrologer and a remarkable individual. Not only was Dee the first to translate the ancient Greek mathematician, Euclid, into English, but he also devised a means of charting the oceans that enabled adventurers like Francis Drake to find his way across to America. He coined the term 'British Empire' as part of his vision of creating a world-wide religion that emphasized the unity of all things and was a close adviser to the Queen. Intriguingly, he may have been a spy for Elizabeth when he travelled to Poland and Prague. I am told that when he wrote to her from such places he signed his letters with a curious combination of symbols: two zeros, sometimes connected by a bridge, implying he was her 'eyes', followed by an elongated 7, which happens to be the alchemical symbol for Mercury. As Mercury was the messenger of the gods, the implication is rather clear. Dee was the Queen's secret and mercurial messenger. And he evidently signed himself 007. Ian Fleming, who created James Bond, adopted that famous moniker after studying Dee's cryptology, although whether he also knew what Dee implied by his use of the 7 is not known. In alchemy it symbolizes the spiritual forces that regulate the cycles of time and human development.

William Shakespeare almost certainly met Dr Dee, and his plays show plenty of evidence of the Platonic thinking that Dee was immersed in. It is widely considered that Shakespeare based his final great character, Prospero, on him, and if he did, the play's central premise would make a lot of sense. Shakespeare pitches Prospero into exile in the wild forest of a far-away island where he is forced to do battle with his demons in a disturbed, spirit-filled arcadia. Only when Prospero heals the disconnection in his own soul can he finally restore harmony to Nature. Then he can make his way home, where he will unify the City State and heal it of its present 'dis-ease'. Like so many of Shakespeare's plays, *The Tempest* charts the course of an individual's deep and personal journey to reconciliation. It is the study of what happens when a man is forced to face the reality of the world and respond from a very deep place to the inner tempest of his soul. It is fascinating how a master of the music of speech like Shakespeare continually relates the heavenly realm to the earthly world in nearly all of his writing. His many sonnets, for instance, are directed towards this relationship. Sonnet 53 explicitly seeks to define the nature of the soul from the very opening words, 'What is your substance, whereof are you made, that millions of strange shadows on you tend?' Most of the sonnets, though, begin in the temporal world and only then draw us towards the realm of the eternal. One of the best known, for instance, begins, 'Shall I compare thee to a Summer's day?', but, passing from the world of time and place where 'rough winds do shake the darling buds of May, and Summer's lease hath all too short a date', we end up in the higher realm of the soul, 'when in eternal lines to time thou grow'st … so long lives this, and this gives life to thee'. This is often read to mean that the memory of the loved one will live on, but if you were to consider these lines in light of the Platonic view of the world that fired so many of the great thinkers of Shakespeare's day, not least John Dee, it is clear that the playwright is saying here that true beauty belongs to the eternal world, the home of the immortal soul. Shakespeare, wielding his pen so brilliantly, conveys the wisdom of the ages in traditional form using stories with many mythological levels, much of which is lost to the modern theatre-goer because we are no longer taught to see our relationship with the world and reality in this way.

I trust the point I have been trying to make by taking this journey through the history of ancient thought is clear. Tracing this golden thread of wisdom and the inner need to maintain harmony in the world and within ourselves demonstrates how beautifully this principle has been woven into the fabric of Western civilization. Clearly the idea is not some wishy-washy, New Age invention of the late twentieth century. Far from it. It is a very precise principle indeed, acknowledged as central by some of the greatest thinkers the world has ever seen. From Pythagoras and Plato to Shakespeare and Ficino, from Giorgione, Bach and Handel to Wordsworth, Poussin and Blake, all of these great artists were very clear that there is a harmony to the world that must be maintained. What is more, the difference we tend to see between the outside, material world and what we might think of as our own personal, mental space within is in fact an artificial distinction. We experience them both as a whole and therefore the balance we achieve *within* us dictates how balanced our behaviour will be without. This is why the ancients considered humanity to be what they called a 'microcosm' of the macrocosm. They saw no separation between Man and Nature and no separation between the natural world and God. Religion and science, mind and matter were all part of one living, conscious whole, with every part of the living world made up of the whole of the universe. They saw people as finite beings contained by an infinitude and reflecting the proportions and purpose of creation. The Stoics stressed the relationship in the idea that 'the proper aim of Man is to live in agreement with Nature. To live in agreement with Nature is to live virtuously, and to live virtuously is to live happily.' But we should be careful here not to see this as a kind of one-way reflection. The process is two-way. Consider the opening lines of the first stanza of the Buddha's great teaching, the *Dhammapada,* particularly as we move into the next chapter and look at what happened beyond this period in the seventeenth century with the advent of the Scientific Revolution and, eventually, the Age of Reason and twentieth-century Modernism. As the Buddha says, 'with our thoughts we make the world'.

The way of patterns

This is why understanding the patterning of Nature seems to me to be so important. It is not just a loose or random collection of patterns. Patterns that are not connected do not make a system. What matters is the quality and nature of the connections *between* the patterns, because these determine whether the collection produces a language or just a confusing babble. In other words, the relationship that all things have with each other is paramount. This is the mistake I believe was made by the avant-garde architects of the Modernist movement and their more recent offspring, who still put their faith in the construction of buildings that deliberately abandon the grammar of harmony.

My concerns about architecture are well known, but the reason I am so concerned about it is because it affects us all. I have never made it my habit to go around criticizing the artwork that people choose to put on their walls, even if I do not like it. After all, that is their business. But architecture is a different matter. It, in large part, defines the public realm and therefore helps to define us as human beings. It affects our psychological well-being because it can either enhance or detract from a sense of community. I say this because I am convinced we are all profoundly influenced by the presence or the absence of beauty and harmony in the places we build and have our being. How a public building looks very much reflects the way we collectively look at and treat the world.

The Union and Pakistani flags at Bradford Town City Hall. This is one of the buildings that children, in a study prepared for my visit, said they preferred over the more modern pieces of architecture.

I remember vividly some years ago visiting a fairly depressed part of Bradford in West Yorkshire and meeting a group of teenagers in a new community centre. They showed me the results of a project they had completed in which they identified and photographed all the local buildings that they most liked and those that they disliked. To my fascination, all the buildings they most disliked were built in the 1960s and 1970s of concrete, steel and glass and all the ones they liked were the few remaining, older buildings, like the town hall,

THE GOLDEN THREAD 135

The car park that towers above Gateshead, England, is an example of Brutalist architecture. Made of raw concrete, its blunt and inhuman proportions have no resonance with Nature's patterns, which are innate in us, and so it jars the senses and offends the skyline, solving the problem of what to do with cars but ignoring the well-being of people who live and work within the mood cast by its shadow.

church and library, together with the small area which had a pond and trees. When I talked with them about this they were unaware of their reasons why, but it seemed to me they were responding subconsciously to that inner, natural language of patterning I have been describing here that is so clearly reflected in the older buildings.

The kind of modern architecture they disliked and that I have been so exercised about is the kind that clashes hideously with this patterning that lies within us. In the most extreme cases – the buildings, for instance, designed by those architects who were even proud to call themselves 'Brutalists' – such edifices seem deliberately to summon up chaos rather than conjure harmony. They actively deconstruct this patterning and this seriously affects our psychological equilibrium. If these buildings were not so disturbing their architects might choose to live in them themselves, but what has always intrigued me is that so few of them do! You will find that many architects prefer to live – and often work – in attractive old buildings, very often situated in the few remaining conservation areas that have somehow survived in our towns and cities. This certainly seems to apply to nearly all of the winners of the renowned Pritzka Prize for Architecture.

I cannot help feeling that much of this disruptive thinking could be stemmed if the timeless principles that make up the grammar of harmony were taught more explicitly in schools. As it is, their studies of modern art seem to leave children with the idea that anything goes. I come across it all the time: a general incapacity to explain why some of the most extreme examples of modern architecture do not feel as welcoming as they should. They may be 'clever' and eye-catching, but they do not nurture a sense of well-being. The reason, I would suggest, is that the collection of patterns they are constructed from do not 'map' the patterns of life within us; they do not harmonize with the natural patterns we have explored here. Whether it be a street or a square, a single building, or the decoration on its walls, if it is to produce that all-important sense of well-being, then the two sets of patterns must mirror. The more they mirror, the more effective they are found to be.

Soon we will trace the rise of Modernism, which did away with this understanding as it came to dominate very much more than our city skylines in the twentieth century, but I mention it here because we have now arrived at that crucial period in European history in the sixteenth and seventeenth centuries that came to define so much of the outlook we have today.

When I first began to study the way the grammar of harmony works in order to apply it to a better way of planning urban environments, I was introduced to the work of the architect, Christopher Alexander. In his inspiring book *A Timeless Way of Building* he quotes an observation from a beautiful Chinese guide to painting that appeared right in the middle of this period in history. It was first published in 1697 and was called *The Mustard Seed Garden*. The anonymous author describes how he became aware that his search for a way of painting had actually been a search for what every other painter has always sought. He thought he was seeking his 'style' of painting, when in fact he was trying to find the central 'way' of painting that thousands of others, just like him, have found throughout history. The painter makes the point that the more an artist understands painting, the more he recognizes that the art of painting is essentially one way – it is a single quest to find the underlying grammar of harmony. It is embedded in timeless principles that are waiting to be discovered and then fashioned in a contemporary way by each new generation. This is because, he tells us, the central way is connected to the very nature of painting. Style is ultimately a meaningless word. 'What we see as style is nothing but another individual effort to penetrate the central secret of painting.'

Christopher Alexander demonstrates that the same truth applies when it comes to building towns and cities. There are, he says, many historic 'styles' of building, but they all have a quality in common, discovered by those who applied the geometry we have been exploring. It is the secret at the heart of all

OVERLEAF: *The pull of the Moon is considerable. Not only does it move tides twice a day, it pulls on the Earth. Many gardeners and farmers are rediscovering the benefit of planting according to its phases, part of a profound knowledge neglected by modern techniques.*

Florence, Italy. A city that grew organically. There is no zoning. It is an integrated complex of streets, alleyways and piazzas where work is done, lives are lived and children play, learning from being amid the work being done by their parents. It is a system of patterns, interdependent at many levels.

architecture. 'The principles that make a building good follow directly from the nature of human beings and the laws of nature, and any person who penetrates these laws will come closer and closer to this great tradition which Man has sought over and over again and comes always to the same conclusion.'

Alexander was very much a part of what became known as the Oregon experiment, which developed his pattern language with the aim of re-anchoring contemporary architecture to traditional architecture. He observed that medieval cities are attractive because they are harmonious. They embody a sense of unity in their overall design. When he began studying these patterns he found that they are just as visible in traditional villages in Malawi as they are in the precincts of Renaissance Italian cities and he concluded that they are 'classic patterns' simply because they have been tried and tested over so many generations. They have always been found to be both practical and pleasing.

I have attempted to demonstrate here that the ancients' grasp of the geometry of the cosmos dictated the designs of their stone circles and pyramids and, in more recent times, the fabulous treasures of the world's current sacred traditions, but also that this cosmic geometry underlies the structure of matter, the biological development of plants, the way in which animals organize their communities, the orbits of the planets and even key astronomical cycles of time. That the great civilizations of the past understood such intricacies and yet our own, so-called sophisticated world seems to have lost this degree of insight and level of knowledge says something rather damning about the breadth and depth of our education – in simple ways as well as more learned ones. How many children, for instance, know which phase the Moon is in today? Or, for that matter, how many teachers – or even farmers, who once would have planned their planting according to the phases of the Moon, know that stronger and healthier crops will result if they are planted at a particular time? We may have become more technically sophisticated, but the Moon has lost none of its gravitational pull on the Earth, so nothing, in essence, has actually changed. Only our perspective and priorities.

The golden thread of inner learning has undoubtedly grown weaker as the West's emphasis on the outer world has become greater. The depth of knowledge yielded when Nature was studied 'as Nature' became depleted as she began to be studied as a machine. What is more, the new fashion for humanism at the end of the Renaissance, plus Luther's challenge to Catholicism that brought about the Reformation, also began to suppress the traditional symbolism that emphasized the role of the feminine principle in the world. We need now to explore how these influences began a process of spiritual asset–stripping that has now eroded almost completely humanity's former insight and wisdom as well as much of the balance that once shaped a very different perspective of the life we inhabit. This erosion is accepted by the prevailing attitude of our day, exemplified by the way our Media portray progress and advancement, endlessly suggesting that it is perfectly possible to live in some sort of shiny, synthetic bubble of convenience so long as we ignore or at least brush under the carpet the corrosive effects that all this is having on the Earth and its essential life-support systems.

Over the many years I have been considering these matters and tracing the causes of this somewhat paradoxical outlook, time and again I have come to the uncomfortable conclusion that the decline in our understanding of traditional wisdom, and its complete eradication from so many areas of life, have not been an entirely natural consequence of the evolution of ideas or the freedom offered by our advances in technology and scientific learning. It seems to some observers that in part its destruction has been designed. Fashionable ideas, clever ways of manipulating public opinion and specific modes of thinking have all altered our daily outlook on life. They persuade us that it is possible to accommodate the excessive costs of our pursuit of unlimited economic growth by maintaining what has become a widespread disconnection from Nature. How this disconnection came about is what we turn to now.

The Age of
Disconnection 4

There are two sciences: the science of manipulation and the science of understanding.

FRITZ SCHUMACHER

The journey through the history of ideas and discoveries is a fascinating one. Like a river that courses through different landscapes, our scientific endeavours have found a way round many difficult obstacles, but just as a river shapes the topography of the countryside that it flows through, so these solutions have shaped our collective mental landscape. It is pointless to accuse a river of taking a wrong turn or choosing a particular course – rivers simply find the easiest way to the sea – and the history of ideas has followed much the same principle. An idea was born, experiment followed and the success of a project then led others to follow suit until progress took a certain direction. Then, at some point, the mood of thinking began to coalesce around this approach. Historians of ideas often report that, as this happens, the original, fluid observations made by the pioneer get exaggerated so that when principles become fixed they are fixed according to those exaggerations. And so a particular way of looking at the world that was never quite what was originally intended becomes more concrete and the pressure to conform more overt. This happened to the way scientific thinking developed in the years that followed the fast-moving world we are about to plunge into. What has ended up today as a very rigid, mechanistic way of approaching nearly every aspect of life was not necessarily the one pursued by those who directed the process of scientific discovery in Europe from the seventeenth century onwards. So there is a need, it seems to me, to review how this came about and, perhaps, to compare the modern impulse that drives the 'exciting' and 'original' new ideas of today's progress with the perspective that was there at the origin.

In one sense I have myself been a victim of this tendency to exaggerate. I have found that if you have the temerity to question conventional wisdom in science, for instance, then you end up being labelled as anti-science, when in fact I have never been so. What I am happy to be considered is 'anti' the *kind* of science that fails to see the whole picture; the kind of science that, because of the particular course it has taken, has eliminated the commonsense understanding of our interconnectedness with Nature and the realm beyond the material. Once upon a time I found this was a particularly lonely position to take but now, I

PREVIOUS SPREAD: *Two instruments that looked into the heavens and forged the future. Galileo's telescope beneath the one invented 60 years later by Sir Isaac Newton, the world's first reflecting telescope. Galileo gradually improved the power of his telescope, grinding lenses himself. Newton also conducted experiments into perfecting lenses and in so doing demonstrated that refracting white light produces the colours of the rainbow.*

LEFT: *Plato 428–347BC held that the whole universe is a divine drama and that philosophy should be seen as a way of inner enlightenment. In this way the Platonic tradition, which informed classical learning, considered all education to be a reminiscence of immutable and eternal ideas.*

am pleased to say, there are an increasing number of scientists who also take this view and are working to widen the parameters again, not least because discoveries in things like quantum physics and in consciousness studies are kicking over the traces of the mechanistic view. This view has been the entrenched position of science for the past four hundred years. Even so, there is a long way to go, particularly when much of science is funded by large corporations with a vested interest in a particular outcome. Science has assumed such authority that its particular ruling on a question can mean vast profits for the companies who fund the science and therein lies a problem which is all too obvious.

Understanding how this mechanistic mindset has come to be so dominant is necessary if we hope to be able to identify when it bites too hard on our perception of things. It will also demonstrate what is excluded when it comes into play. So this section of the book charts the way in which the mood was set as the Scientific Revolution launched the modern world and how its specific language became the only language by which we are allowed to consider the world, particularly since the emphasis on Modernism became so comprehensive in the twentieth century. The combination of these and other related influences has carved the mental landscape we find ourselves in today. Although we seem not to think so, it is a narrow gorge that obscures the view of the wider horizon, and my aim here is to lift our perspective so that the full picture is revealed. If we can do this we will better understand where the fractures in our perspective occurred and why. If we can heal those fractures, surely we may then be better placed to heal the Earth.

There are many factors that have shaped the modern Western attitude to Nature, but if I were to put my finger on the biggest ones, I would point to three: the fascinating changes in human perception caused by the Scientific Revolution of the seventeenth century, the impact on our outlook of the Industrial Revolution of the eighteenth century and the deliberate demolition job carried out on traditional culture by what became known as 'Modernism' in the twentieth century. I want to explore the three of them here as we pick up from the last chapter at the point in the story where Europe began to turn away from its faith in religion and towards our present faith in science and reason.

Revolution and reduction

Seventeenth century Europe was rarely a peaceful time. There were many great changes afoot as the old order began to be challenged on many fronts by the new. Central to the politics of the time was the increasingly desperate struggle by the Roman Catholic Church to maintain its authority in the face of two

great challenges: the growing trend towards humanism and the Protestantism that was questioning the supreme power invested in the Pope. The latter in particular created a tinder-box that ignited into war several times during that period. It is hard to say which side was the victor in the religious wars of the seventeenth century. It is far easier to decide whether belief or reason won the struggle in the wider battle for hearts and minds. The individual's submission to the authority of the Divine began to weaken in the Western world as the freedom of self-reliance began to take centre stage.

That transformation was due in no small way to the Scientific Revolution, which established the authority of a mechanistic approach to thinking, often called 'reductionism'. It was reductive because it involved separating out the elements that made up an organism, reducing or atomizing the organism to understand how it worked. It is this reductive approach to studying the world that, in my view, came to dominate thinking beyond the realms of the scientific laboratory. It persuades us now to see the whole of the world as one of cold and separated utility. But what made this possible? How did it come to be? After all, the Scientific Revolution did not happen overnight.

There have, of course, been huge benefits from the path we have taken, and these were clearly a driving force that accelerated progress to its present speed today, but I have long felt sure that, quite apart from the sheer eagerness displayed by the great scientific pioneers like Galileo or Francis Bacon to understand the wonders of the world, there was a deeper reason why, generally, people felt more and more comfortable with the idea of exploiting Nature by mastering her ways. So the discovery I came across some years ago that there was, indeed, a possible cause of this change in thinking was as encouraging as it was revealing. I do not intend to turn this into some sort of philosophical exposition – what I would like to do is explore the solutions we might adopt in today's modern world – but I do feel it is important to expose a certain fact of history that is little considered any more, if only so that it may be more widely considered. When I was first shown it, by Dr Joseph Milne of the University of Kent, I was certainly astonished because it seemed to me to be a missing piece of the jigsaw and confirms what I have always felt intuitively to be true. So, if this is to be a complete explanation of how I have tried to put the best of ancient learning back into practical action for the future, it would be wrong of me to exclude this discovery at this point.

I found it a fascinating revelation to be taken back to the thirteenth century and to look, not so much at what scholars of the time were writing, but how they saw the world. There were few universities then but in one of them, in Paris, sat the great thinker of his day, Thomas Aquinas. At the time, the Church was still the undisputed authority on all matters, including science, so it fell to someone like Aquinas to decide the big questions of his day. It was his task, for example, to judge Aristotle's fate. The work of the ancient Greek philosopher had recently come to light via the Arab world of Spain and Aquinas had to advise whether Aristotle should be considered a friend or an enemy of the Church. Aquinas operated within a traditional view of the universe. If you were to read his work you would find that he describes something he calls the 'Eternal Law'. What he means by this is the law that exists in the mind of God. Aquinas saw no separation between creation and God and he taught that we should

experience the world very much from the inside out. In other words, the prevailing attitude, certainly as Aquinas began his studies, was that the Creator was not separate from His creation. Instead, divinity was considered to be innate in the world and in us. The natural world itself was an expression of this sacred presence and in such a created unity, humanity had an active role as participant. This is the view that still prevails in many indigenous, primary cultures of the world, as we shall see later. Many times the Qu'ran describes the natural world as the handiwork of a unitary, benevolent power and very explicitly points to Nature possessing an 'intelligibility.' It finds no separation between Man and Nature precisely because there is no separation between the natural world and

God. It, too, offers a completely integrated view of the Universe where religion and science, mind and matter are all part of one living, conscious whole. This is the relationship the Stoics of Ancient Greece responded to with their idea that 'right knowledge' can only come about by living in agreement with Nature, where there is a correspondence or a sympathy between the truth of things, thought and action. They held that it is our duty to achieve an attunement between human nature and the greater scheme of the Cosmos. The Daoism of China or the Vedic tradition of India also operate from this point of view. In fact what the West calls 'Hinduism' is perhaps better described as 'Sanatana Dharma', which translates as 'the ever-living law' – very much the Eternal Law Aquinas talks of. It was this image of 'God' that Aquinas inherited; that 'God' was the principle at the heart of the unfolding universe – what the Welsh poet, Dylan Thomas, called 'the force that through the green fuse drives the flower.'

This sense of the unity at the heart of all things, as we have seen, is central to traditional Islam, and remains a mainstay of great sacred traditions of both the East and the West.

Bearing all this in mind, then, what I found particularly intriguing when I was guided through this medieval view of things by Dr Milne was what appears to have happened to that prevailing mood of thought during the thirteenth century. For a variety of complicated political and theological reasons, a different definition of God began to emerge. Slowly but surely God began to be defined as something that lay outside of creation and was separate from Nature and, as that happened, so Nature itself came to be seen more and more as an unpredictable force; something likely to be unruly, without inherent order and capable of going its own, sometimes dark, way. It could do this regardless of the so-called Divine Will because the Will of God had become external to the created world. That is an extraordinary shift in the collective perception of experience.

The first evidence of this intriguing shift comes in the way education began to break apart. Aquinas's approach to learning at the University of Paris had been the traditional, classical one, which aimed to arrive at an integrated knowledge of reality *as a whole* – one that was apparent in the outside world, but had its roots on the inner level. In other words, it incorporated what was known scientifically or empirically with what was understood philosophically and also sensed spiritually. But as this fundamental shift began to take hold, so each discipline began to take its own separate course, and so the integration of scholarship ceased to be the central aim of learning.

This may still seem terribly esoteric, but I have to say I find it very revealing. Clearly this was a highly significant shift in the collective perception of Western

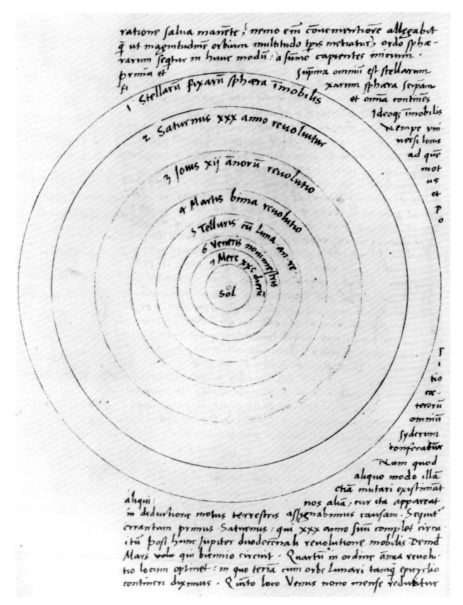

ratione salua manente, nemo enim conuenientiore allegabit q̃ ut magnitudine orbium multitudo tp̃s metiatur; ordo sphæ-rarum sequitur in hunc modũ: a summo capientes initium.

1. Stellarũ fixarũ sphæra imobilis

Supma omniũ est stellarum xarum sphæra seipam et omnia continens Ideoq̃ imobilis nempe universi locus ad quẽ motus et po

2. Saturnus xxx anno reuoluitur

3. Iouis xij ãnorũ reuolutio

4. Martis bima reuolutio

5. Telluris cũ Lima ãnuã reuolutio
6. Veneris nonimestris
7. Merc xxc dnii

Sol

sitio reliquorũ omniũ syderum comparatur Nam quod aliquo modo illa etiã mutari existimat

aliqui:
nos alia, cur ita apparreat in deductione motus terrestris assignabimus causam. Sequnt' errantium primus Saturnus: qui xxx anno suũ complet circu-itũ post hunc Iupiter duodecimali reuolutione mobilis Deind Mars volu qui biennio circuit. Quartũ in ordine ãnua reuolu-tio locum optinet: in quo terra cum orbe Lunari tamq̃ epicyclio continori dyximus. Quinto loco Venus nono mense reduxitur

A page from the work of Nicolaus Copernicus showing the position of planets in relation to the sun. His book On the Revolutions of the Celestial Spheres was published in 1534. He only just saw it go into print. It is said that as he died, a copy of it was placed in his hands. He was buried in a heretic's grave and only given a proper funeral in May 2010.

thought. In time it framed the outlook that allowed science to make its clean break from religion and forge ahead towards modernity. It effectively shattered the organic unity of reality, which could be traced back to Plato and Pythagoras and, before them, the Egyptians and the start of the Vedic tradition in India. At the heart of things, within a very short space of time, that all-important, timeless principle of *participation* in the 'being' of things was eliminated from mainstream Western thinking. Or, to put it more graphically, with God separate from His Creation, humanity likewise became separate from Nature.

Nature began to be seen as something outside of us. We were still a part of creation as other things were, but we were no longer creation itself. Leading thinkers began to stress the role of humanity as the instrument of the Will of God, an instrument that was free to pursue a mastery of the will *over* other things in Nature.

This seems to me to have been the great fracture in thinking that prepared the ground for all that was to come in the turbulent world of the seventeenth century. The seeds took their time to germinate, but in the end they sprouted and so our modern culture was born and has since created a general attitude in society that has little problem with the view that we live in an inert, purposeless, mechanistic universe. As this mechanistic world view grew and, eventually, as industrialism took over, so Nature was more and more reduced to what it is today: raw material. Raw material is a functional thing, but it can only have a function when it is removed from the ecosystem that supports it, put into a factory and turned into something else that is sold as a product. I can hear people saying now, 'And what is wrong with that?' The answer is, nothing at all if we pay attention to the whole picture. As it is, we only concentrate on the outcome, on what rolls off the production line. We do not take into account in an immediate way the impact that this has had on the entire system – which, by the way, also includes us.

The great divorce

Galileo Galilei had been a professor of Mathematics at Padua University when he first made his name by taking a toy invented in Flanders called a spyglass and turning it into a powerful tool for trade. He realized that in the febrile commercial world of Venice a broker in the hurly-burly of the markets of the Rialto could make a lot of money if he could spot and discover what was on board a heavily laden merchant ship when it was still some two hours from port. But it was when Galileo increased the magnitude of his new 'telescope' and turned it to look at the stars above the Venetian lagoon that he helped to set science upon its singular journey into the unknown.

His discoveries of Jupiter's four satellites and his mapping of the surface of the Moon sent shock waves of excitement throughout Europe. Here was final proof of the theory, suggested by Nicolaus Copernicus a generation before, that the Earth was not the centre of the universe. It was in fact like all the other wandering planets, a small satellite of the Sun.

Galileo paid heavily for his adventures in space. He was tried by the Church for heresy in 1633 and, although he escaped the fiery fate of many other con-

The cover of a history of the Royal Society of London, founded in 1660 and made Royal by the new King Charles II. His bust sits between the first president and Sir Francis Bacon, who, much revered for his analysis of experience 'by mechanical means so as to arrive at true conclusions', was dead by the time it was founded.

demned heretics, he was sentenced to house arrest for the rest of his life, an event that has come to mark the famous split in the Renaissance between faith and reason.

That split was also being forged, perhaps unwittingly, by Sir Francis Bacon in England, who published a highly influential book on science in 1620 called the *Novum Organum*. The 'new instrument' this title referred to was the process of reduction. In his book, Bacon describes an enquiry into the workings of Nature as being a process that should pass through 'progressive stages of certainty'. This set in train the way in which science breaks 'being' down into its quantifiable and measurable parts. Organisms are fixed or pinned, clamped,

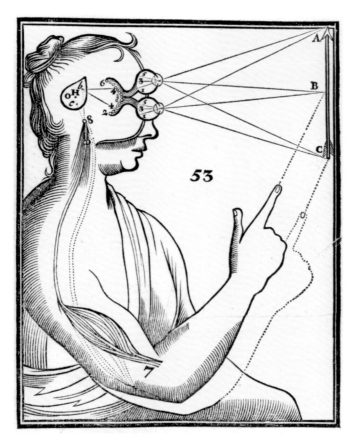

53

A drawing of the way we perceive the world, according to Descartes in 1686. He arrived at his famous conclusion 'I think, therefore I am' by eliminating anything of which there could be doubt, seemingly ignoring that the 'beingness' of all things must first be there for it to be eliminated. Hence the subtle wisdom of the ancient Vedic proposition of India, 'Ho Sum' which means 'it is, therefore I am.'

pressed or pulped to extract yet smaller parts for analysis. These days they can be very small indeed – molecules and genes – even though, as we shall discover later, this 'fractionizing' of a creature and the 'atomizing' of reality has been revealed by science itself to be a limiting method of investigation. Quantum physics, for instance, has blown some interesting holes in the classical mechanics of Newton's theory. This study of matter at the tiniest of levels reveals that electrons can only form certain patterns as they flit about in atoms. The structure of atoms – which dictates the whole of chemistry – and the structure of the smallest elementary particles that make up the structure of the atom, both conform to certain rigid principles of mathematics. At this level such structures have been described as 'perfection'. They create a dance of form and matter, a dance that has been described to me as geometry itself, involving trillions of particles that constantly flip between existence and non-existence on so small a time scale that it happens between what physicists actually call 'a moment'. This is all hard to understand, but it suggests that the fabric of space is an astonishingly dynamic thing. Classical mechanics, which assumes that matter is continuous, could not explain such perfection and certainly cannot measure it. Interestingly, cutting-edge research into the nature of consciousness still struggles with what it is that makes an organism conscious and gives humanity its sense of being.

This 'sense of being' seems to me to be a crucial element that was sidelined by so-called 'Cartesian' thinking that began its rise to prominence in the middle of the seventeenth century. Named after the mathematician and philosopher René Descartes, the Cartesian approach was to imagine the world as a vast and complex mechanism. If we want to know how an element in Nature works, said Descartes, it can help to see that the organism is a machine. He suggested that animals did not have feelings because they did not have souls and in this way they can be seen simply as mechanisms to be wound up or wound down.

Despite being a religious man, Descartes still felt it possible for humanity to manipulate these machines and exercise unbridled control over them, which is still very much the convention today.

I trust it is clear that I am not suggesting remarkable people like Bacon, Galileo and Descartes set out to destroy the world. None of them was an atheist. They all acknowledged the reality of mind and of spirit, but they did separate these elements of their experience from their scientific method of exploration, leaving these to the philosophers and theologians, and this became the set pattern. As people looked back on their approach and defined the route by which they had achieved their great discoveries, so the world became fixed on a particular method of exploration that excluded the other levels of thought and the non-material aspects of being and in this way empirical science and the mechanistic view of the universe was pushed more and more to the fore.

This 'sense of being' seems to me to be a crucial element that was sidelined by so-called 'Cartesian' thinking that began its rise to prominence in the middle of the seventeenth century.

Certainly by the end of that volatile and bloody period in European history during the seventeenth century a messy divorce between science and religion had occurred. What had been a purposeful and unified understanding of Nature was replaced by a purely scientific, mechanistic conception of Nature. It is interesting that the Catholic Church has never accepted this reductive view of creation but, even so, the Christian Church in general has struggled to keep science within the larger unified view of Nature, not least because the majority of Christians have been swept along by the scientific view of the natural world regardless of the Christian view. If I may delve a little deeper for just a moment, it seems to me that this happened because confusion exists on both sides. Religion's notion of science is different from science's idea of itself, just as 'scientism', which is what we really mean when we talk of science, has a some-what false notion of religion. Certainly it is the case that the split was established between science and religion *by the scientific community,* but this split now goes much deeper. It is not merely one that divides two institutions, religion and science: it is a split dividing the psyche in many societies, and it is therefore present in most individuals.

The outcome of this divorce is that from the seventeenth century onwards science was able to view Nature in terms of its mechanisms. Nature became completely objectified. 'She' became 'it'. Nature was deemed by science to be inanimate, unconscious and mechanistic. Intelligence and purpose were not to be found in Nature, but outside it, the property of some sort of transcendent God which religion could deal with on its own. Meanwhile, humanity came to be seen as having the right – a human right, that is – to explore, manipulate and exploit every element of the natural world for the betterment of mankind. What is more, as time went by, scientific rationalism began to expand its profile so that today, as I have already mentioned, it decides the ethics of a debate in questions of, say, genetic manipulation. It certainly has a much more authoritative voice than philosophy, ethics and religion.

By the end of the eighteenth century the idea that the universe was a living presence, that it had purpose and was endowed with an inherent intelligence, had become, as it remains very much today, an inconceivable notion. Gone was God, the constant sacred presence that participated in the being of things and, in His place, an external, separate arbitrary Will that imposed itself upon Nature, with humankind acting as its instrument. This arbitrary Will is the God in the sky, outside the world of things but connected to them through us His instruments. By then, human wilfulness had adopted the name of 'Reason' and Nature was being plundered and spoiled in the name of Reason. There was nothing sacred about it at all. Nature had become nothing more than a great opportunity for experimentation and the supplier of natural resources. Take this rather unnerving observation about how to deal with pests from something as apparently benign as *A Treatise on the Culture and Management of Fruit Trees*, compiled in 1803 by an Englishman by the name of William Forsyth, who wrote: 'It would be of great service to get acquainted as much as possible with the economy and natural history of these insects as we might thereby be enabled to find out the most certain method of destroying them.'

By that time this detached view of the world around us had become very dominant in many significant fields of endeavour. I have been particularly fascinated by the extensive research done of late by Dr Milne on how attitudes to the law changed in Europe during the seventeenth and eighteenth centuries. Matters of law have always attracted the greatest thinkers of the day – being so complicated perhaps they have to – but notice the striking difference between the way modern thinkers began to approach the law once they loosened the ties to classical thinking and the way in which human nature had been conceived before that time. The leading thinkers from the mid-seventeenth century onwards saw the law as we tend to see it today. People like Hobbes, Mill, and Rousseau, who helped to develop the concept of modern individualism and whose Utilitarianism eventually gave birth to modern consumerism, all regarded the law as a means of curbing the tendency towards what they called the 'savage inclinations' in human nature. As Dr Milne has shown rather incisively, even though Rousseau in his most famous book, *The Social Contract*, considered Man to be the 'noble savage' he still maintained that there is a conflict between the 'natural' human state and the 'social' one and so, for him, along with his contemporaries, society was seen as 'artificial' with the law being a pragmatic invention that is not grounded in any universal natural order. For Rousseau, society exists to defend against threats and dangers, not for the attainment of the universal good. This was not the view of human nature taken by the

Westfield shopping centre, London. Consumerism is now at the core of Western culture. Shopping defines who we are and is about more than meeting our practical needs – a triumph for sophisticated advertising. But is it possible to inspire a more sustainable way of living?

great thinkers of ancient Greece or Rome. Figures like Plato and Aristotle did not believe human nature was fundamentally savage. They held that it inclined towards the good and true. Aristotle, for instance, believed that we are, first and foremost, 'political' creatures: not just gregarious like other social animals, but creatures who delight in the mutual recognition of universal law, justice and the common good and do so in union with the natural world. For Plato, this union is the source of our sense of law and our capacity to formulate laws through collective understanding. So in this sense, law is the expression of the essential sense of the unity and harmony of Nature. It links contemplation with action. Big and small, the impression is very clear as to what was unwittingly happening. It had not been the plan, but, Western thought and its outlook were laying the foundations of our present Age of Disconnection.

Not that it seemed like that at the time nor for many centuries after. The idea of progress through science and technology was the established mainstream political and economic agenda from the eighteenth century onwards, which, with its emphasis on human rights, the triumph of reason, and the importance of the free market, is known to us today as the Enlightenment. Part and parcel with the progressive movements came the accepted authority of science. I am the first to acknowledge that the Enlightenment caused wonderful things to happen, but I do wish that the champions of mechanistic science would be more prepared than they are to accept it also brought downsides with it.

The industrialization of the mind

The very fact that such a notion can find the credence it does in our world today is down to the influence of those two other related developments: the impact of three centuries of industrialization and the twentieth-century ideology that emphasized the speed and convenience of the machine which became known as 'Modernism'. Perhaps a specific example that affects us all might be the most vivid way of describing how industrialization benefitted from the authority of mechanistic thinking.

Consider the legacy of one Justus von Liebig, who grew up in Germany in the early years of the nineteenth century. One day in 1817 there was a very loud explosion in a quiet street in Darmstadt where he lived. The window frame of his family's house blew out and bounced in a cloud of shattering glass across the road. Justus's father dealt in painter's supplies and domestic chemicals and Justus had become fascinated by them. Ever since a travelling pedlar had sold him a toy torpedo, propelled by a fulminating combination of substances, Justus had been experimenting with the chemicals it contained. To begin with he had taken the torpedo to bits to discover that it was propelled by a mixture of mercury, nitric acid and alcohol but then he set about making his own, much more successful toy torpedoes, which his father had taken to selling in his shop. The trouble was, he could not put these fulminates down and his experiments became ever more dramatic until he finally took a load of them to show his friends – and that is how his school career came to a sudden and rather explosive end.

The Liebigs decided it would be safest to send their obsessive son to live in another town. He was apprenticed to an apothecary and in just ten months Justus had mastered the profession, but he also carried on with his private combustible experiments in his spare time. At the age of 17, and no doubt to the relief of his landlord, Justus Liebig secured himself a place at the University

of Giessen where, by the time he was 21, he had become the youngest professor in the university. At last Liebig could call himself a 'chemist'.

Perhaps because of the volatility of his early studies at home, Liebig appreciated the importance of a disciplined approach in the laboratory. He set about developing a course of study at the university that rapidly became the model for nearly all laboratory instruction, not just at Giessen, but throughout Europe. He adapted a deserted barracks in the town where he had his students prepare organic compounds and carry out investigations into their properties under his direct tutorage. His fame and his new approach to teaching quickly spread and Liebig eventually became the premier teacher of chemistry of his generation. Indeed, his disciplined approach to laboratory technique crossed to America as well as spreading throughout the burgeoning university system in Europe, so much so that if Liebig were alive today he could very well lay claim to having helped produce many of the Nobel Laureates in chemistry and biology who have followed in his footsteps.

Liebig has many claims to fame. In 1865 he founded the Liebig Extract of Meat Company, better known today for its little cubes of beef extract called Oxo. But before he hit on such a popular innovation, in 1840 he published a book called *Organic Chemistry and Its Application to Agriculture and Physiology*. With a title like that it was never destined to become a bestseller, but it was a

Kelp Gathering in 1890 on the coast of County Antrim, Ireland, used as natural fertilizer before artificial ones were developed. Globally we now use over 1 billion tonnes of synthetic nitrogen fertilizers every year, a 20-fold increase on 50 years ago.

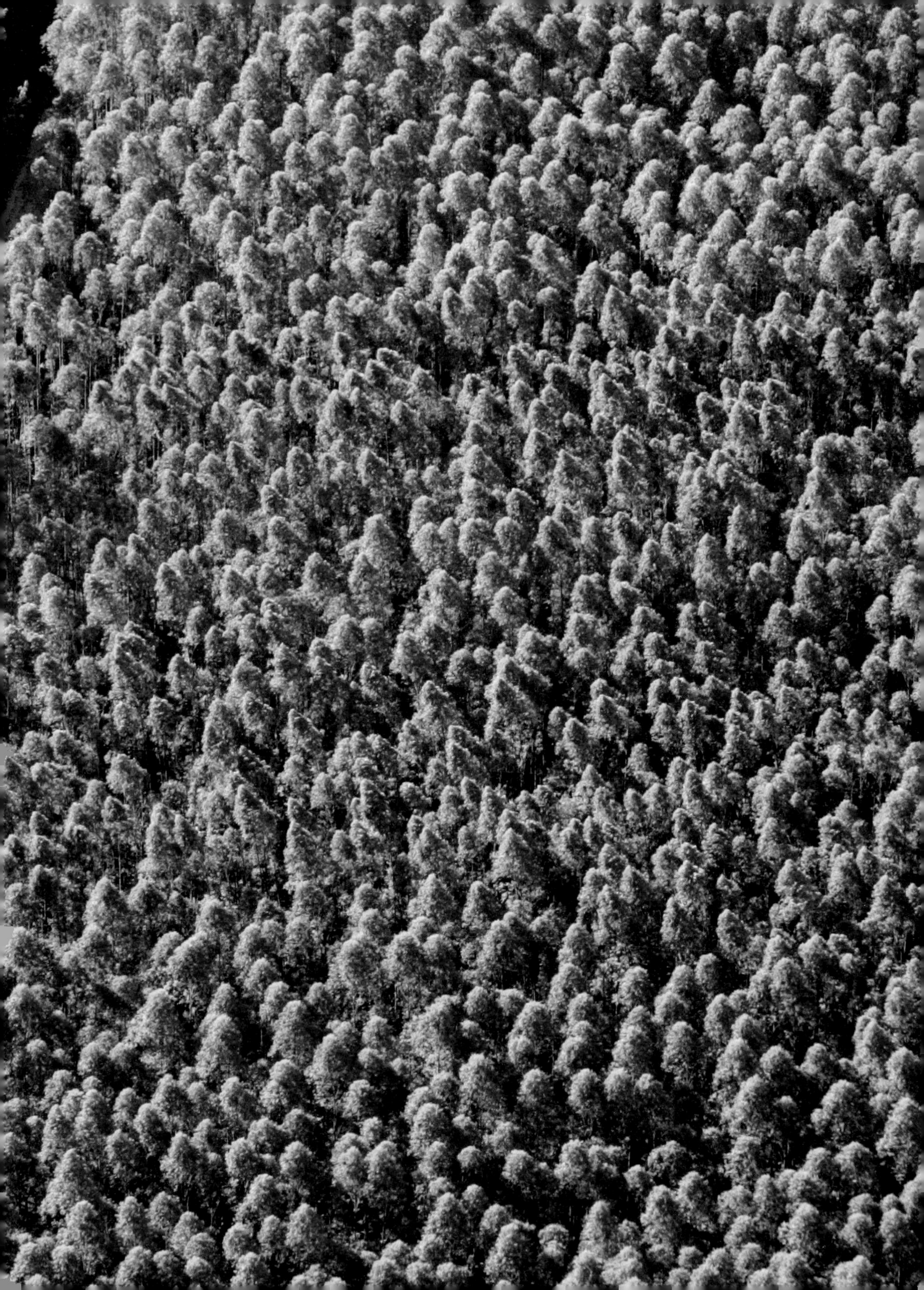

PREVIOUS SPREAD:
*An aerial view of the
Brazilian Amazon
showing the contrast
between a naturally
diverse ecosystem (left)
and the regimented
near-monoculture
of just a few species.
It may still look like
a forest, but a
monoculture does not
support the same
complexity of life.
It can be akin to a
green desert.*

highly influential book, winning Liebig the nickname 'the father of the fertilizer industry'. This was a highly technical early encyclopaedia of chemical farming.

Liebig had set out to define what makes plants grow. He took crops, set fire to them in his laboratory and studied the ash that was left behind to identify the minerals that provide plants with their necessary nourishment. His analysis revealed three: nitrogen, phosphorus and potassium, and this was a substantial discovery. They are the three basic minerals every farmer and gardener knows today as NPK. But that was not the only information to enter the popular domain via Liebig's researches. He also helped to establish the notion in early agricultural chemistry that a plant is little more than a chemical processing factory, turning this vital combination of minerals into energy. Furthermore, by mixing these elements in a solution, Liebig also demonstrated that plants can grow perfectly well without the presence of soil, and that was a very significant development indeed. It set agricultural chemistry on the path that eventually led to the industrialized approach to farming we have today where, in effect, the conditions of Liebig's laboratory have been transferred as much as they can be to the open field. It is an approach based upon the damaging assumption that plants do not have to depend upon the living and other organic material in the soil called humus. This extraordinary substance is the combination of decomposed organic matter and the microbes that do the decomposing, but in conventional farming their function is bypassed and disregarded. The soil is kept artificially fertile with synthetic fertilizers. This leads to many big problems, as mechanistic thinking on its own often does. For one thing, fertilizers do not discriminate. In their concentrated form, minerals like nitrogen encourage plenty of growth in everything they are applied to, so, without some sort of preventative action, the crops that are doused with highly effective synthetic fertilizers become overwhelmed by weeds, which is why conventional farming also depends upon a considerable cocktail of herbicides and pesticides to keep weeds and bugs at bay.

A century after Justus von Liebig conducted his pioneering research, the philosopher Rudolf Steiner was asked to deliver what became a famous set of lectures on the emerging crisis in agriculture. Steiner was quite clear about his view of this development. He described Liebig's approach as taking agriculture out of the realm of life and putting it into the realm of death. Only in the realm of death, said Steiner, does a theory like this work.

For his part, Justus Liebig appears late in his life to have seen the error of his ways. He tried to correct the balance by writing that 'through a higher force at work in living bodies of which inorganic forces are merely the servants, do substances come into being which are endowed with vital qualities'. Unfortunately

by then the enormously profitable bandwagon of chemical farming had already begun to roll and the many commercial entrepreneurs whose substantial incomes now depended upon its acceleration were not going to allow it to slow down. Liebig could not defeat the monster he had helped to create and today we live with the legacy of his pioneering work, whereby the vast majority of the food we eat is produced by a method of farming that has become alarmingly disconnected from the Earth. Instead of growing it through a natural interaction with the complex web of life in the topsoil, farming has made it more reliant upon various combinations of manufactured chemicals. It is a system that is said to have made enormous strides to feed the burgeoning world's population,

but those advances have to be weighed against the huge and unsustainable costs it brings to bear.

Despite Steiner's warning a century ago, together with two world wars and a dramatic acceleration in the world's population throughout the twentieth century, pressure has mounted on agriculture to adopt the clinical efficiency of the factory production process. To meet this demand a new approach was devised in the USA and then rolled out elsewhere in the world, first in Western Europe and then, in the 1960s, in countries like India. This was the so-called 'Green Revolution' I mentioned earlier.

The claim from those Nobel Prize-winners who devised this scheme is that this is the only way to feed the world and yet, in places like India, where there has been a thirtyfold increase in the use of artificial fertilizers since the 1960s, levels of micronutrients in the soil have fallen and continue to fall, with the result that yields grow no bigger. In fact in some cases they have also fallen. The yield for every unit of fertilizer applied in those parts of India where the Green Revolution was imposed most vigorously decreased by two-thirds during the early stages of the roll-out and in more recent years there has been an alarming loss of fertility in the soil, which combines to produce an extremely worrying, diminishing return on a process that is promoted ever more vigorously around the world.

Nature does not make such mistakes. She does not grow just one kind of plant intensively, poisoning all else to make it prosper. Hers is a far from silent Spring. It was the ecologist Evelyn Hutchinson who, nearly fifty years ago, asked the very simple question, 'Why are there so many species in Nature?' Her own answer was that numerous species maximize the chances of creating what she called 'reciprocal relationships'. These relationships enable different organisms to store energy and resources for one another. As she demonstrated, the benefit is that plants tend to be more resistant to the stresses of drought, pests and diseases. By growing just one, genetically uniform plant in an isolated, linear way, we reduce the biodiversity of the entire ecosystem and therefore weaken every plant's robustness. Just as a child who is not exposed to germs will not develop a resistance to infection, so plants grown in this way become less and less able to fight off disease.

It is a staggering fact that in just one pinch of soil there are more microbes than there are people on the planet, and yet we know next to nothing about what these microbes do. Given the numbers involved it is reasonable to claim that, on the microscopic level, that single pinch of soil is itself a complex, inter-related living ecosystem, and yet we soak it in chemicals that we believe, from our great distance, only destroy weeds. What they actually do, of course, is

There are 50 billion chickens in the world. They outnumber every other bird on the planet. A chicken will live for 8 years but reared in windowless rearing sheds on an industrial scale they are ready for slaughter in just 6 weeks. They have no room to exercise, forage for food, dust-bathe or any of the other natural habits of their behaviour.

destroy life and given that we know next to nothing about the myriad functions these billions of microbes perform, it is a breathtaking piece of arrogance that we assume such a cocktail of chemicals does no long-term harm.

Despite what Liebig managed to demonstrate, plants do need the soil to grow. The humus contains a vital, living presence that keeps the fabric of the soil strong and its fertility high. Our modern-day conventional approach is to ignore this. As a result, the soil has suffered a worrying degree of depletion and erosion since these techniques were first introduced.

It is surely obvious how the mechanistic mindset I have described is at work in this process. It is not a process that has anything to do with traditional thinking. But if there is one area of human activity that demonstrates this approach more graphically than any other it is the appalling way in which animals are now reared to produce milk and for meat in industrialized countries. Descartes famously said that animals are mere machines devoid of feeling or sensation, and that seems to remain the main justification for the horrific way in which millions of chickens, beef cattle, and pigs are 'manufactured' to specific weights and standards of meat in so-called developed nations. In countries like the USA and Brazil the process is completely industrialized and very much dependent upon a disconnection from the benevolence of natural systems. Millions of 'farm' animals today never see a farmyard, let alone a field. Pigs and chickens are kept penned up all of their lives in vast, airless sheds where they are fed foodstuffs that Nature never intended them to eat. They are also pumped with growth hormones and kept alive with heavy doses of antibiotics which, of course, help to produce an increasing resistance to antibiotics within the human population. Carrying the excessive weight they do and unable to exercise, these pigs and chickens lose the strength in their legs to stand. Many chickens kept in such conditions die before they ever reach the required weight.

Outside, beef cattle fare no better. They are reared to begin with on intensive farms and then acquire most of their weight when they are moved to the vast feed lots where they rarely see a blade of grass again, far less eat one. They are fed corn, often genetically modified corn, which is artificially cheap to produce because of the subsidies in place, even though it leads to all sorts of illnesses and bacterial problems because cattle are not designed to eat corn – they are, after all, 'herbi'-vores. What is more, they stand all day knee-deep in their own dung and as they are cramped together the chances of disease are heightened.

All of these animals are transported sometimes thousands of miles to the very few vast factory slaughterhouses that now produce most of the meat in industrialized countries. I was astonished to discover that in the United States, a country with a population of around 308 million, there are now just thirteen

slaughterhouses supplying the vast majority of beef to the people of the country. The same extraordinary ratio exists in many countries in Europe. These slaughterhouses are enormous and completely mechanized. The levels of stress the creatures must suffer are unimaginable. These slaughterhouses the size of car factories process tens of thousands of animals every day and with so much of the meat destined for the markets in ground beef or pork and processed chicken, the meat that ends up on the table will never come from just one animal. A single hamburger sold in a fast-food restaurant, for instance, will contain the meat of thousands of different cattle, much of it treated with ammonia to cleanse it of traces of fatal bugs like *E. coli*. This processed meat will also have travelled many more miles, thousands of them, to reach the next stage in the processing phase before eventually arriving on the bright and shiny shelves of the supermarket or served in its polystyrene box over the counter.

This system is vastly expensive to keep moving and just as hugely thirsty of water and oil as the arable process, but the predominant agenda of achieving progress through science and technology has persuaded us that Nature's capital is unlimited and that it is all there for us to manipulate and exploit, however extreme the techniques and destruction may be.

Far from this approach being limited to the USA, where it was first developed, the same intensive processes are now being rolled out across Western and Eastern Europe. As this book goes to press I gather there is a group of farmers in the East Midlands of England planning a vast industrialized dairy operation that will involve 8,000–9,000 cows being incarcerated in a kind of factory all of their lives. The backers argue that these creatures will be free to roam in open-sided sheds, but that is not the same as outside. I have no doubt that they will be fed concentrates and will last for only one or maybe two periods when they are producing milk – a 'lactation' – before being declared redundant, all done because of 'efficiency' and economy of scale. This is the ultimate, disconnected horror of the modern world view and I consider it an analogy of all that is fundamentally wrong with the Modernist ideology that has finally reduced everything to a mechanistic, industrial process, even animals themselves. And the awful, obscene irony is that such an approach will merely compound the problems of disease by deliberately ignoring Nature's economy.

The values of our culture very clearly dismiss any feelings of unease we may harbour that such an attitude may be flawed. It is, as I said earlier, a world of cold and separated utility and we accept it because we accept what our science tells us. It produces jobs. It produces an attractive bottom line. We are told also that it is safe, that the chemicals involved do not cause irreparable damage to the fabric of the soil, nor to us; that it is perfectly acceptable to mess with the genetic make-up of the most important of all substances on Earth – grasses and their grain. We accept all this without question. We believe what our science tells us to believe, even though the majority of us do not understand the science at all.

Machines for living in

In no small way this is a result of the degree of disconnection from the cyclical patterns and economics of Nature that passes for everyday life in the twenty-first century, but this 'everyday life' is very much the product of an ideology that became ever more prevalent during the course of the twentieth century – an ideology that, in my opinion, is seriously in need of a review, followed by reform if not rejection. That ideology is Modernism.

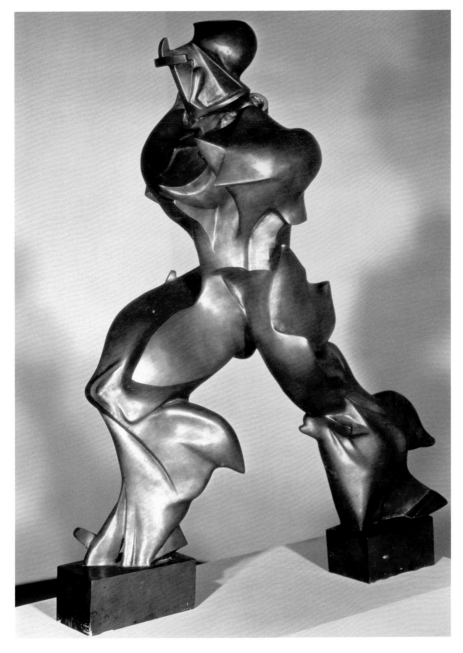

The bronze Futurist sculpture Unique Forms of Continuity in Space created in 1913 by Umberto Boccioni celebrating speed and the forceful dynamism of the machine age. A prophetic vision, indeed, of the merging of Man and Machine that would come in the 20th Century.

In truth, 'Modernism' is a term that refers to what was a loose collection of ideas, promoted by more than one movement that emerged in Northern Europe at the time of the First World War. Essentially, all these movements rejected both history and Nature as central sources for design. Instead they focussed on technology and abstraction, legitimizing a kind of mythology of industrialization. Unsurprisingly, it was an urban movement, where it was more

The vast secretariat building, home to the Punjabi government in India, designed by Le Corbusier and employing the unfinished 'raw concrete' look with which brutalism became so synonymous. It is every inch a Modernist design and presents a very inhuman image of government.

possible to consider the natural world in an abstract way. It flourished in Berlin, Moscow and Paris, in Rome and Prague and eventually in New York, and its genuine aim was to create a better world using technology as the key to social improvement. In Italy the Futurists, for example, based their entire vision of utopia on the wonders of technology. Their spokesman, Marinetti, called on his countrymen to cast off the second-hand clothes they had been wearing for so long and create a new art forged from the beauty of speed and, believe it or not, a glorification of war. Marinetti had a genius for publicity and was one of the first Modernists to realize the central significance of the Media to convert people to his way of thinking. He published a manifesto of his ideas in Paris before he had even produced any art to express his ideas. He wrote, 'We declare that the splendour of the world has been enriched by a new beauty: the beauty of speed. A racing automobile with its bonnet adorned with great tubes like serpents with explosive breath … a roaring motor car which seems to run on machine-gun fire, is more beautiful than [the sculpture of] the Victory of Samothrace.' The central symbol of Marinetti's aspirations was clear. Like others in the Modernist movement, he envisaged a world in which beauty was defined in terms of the machine, with everything recreated on machine-like principles, from clothing to music to theatre. The Futurists wanted to strip away all of the clutter of traditional thinking and create new methods to transform society. Even the human form was depicted as a machine-like being, although I suspect the sculptor, Boccioni, could not have imagined when he produced his famous sculpture that he was being quite as prophetic as he turned out to be.

For the early Modernists, the machine combined the possibility of convenience with simplicity and efficiency, all of which seemed possible in the first decades of the 1900s as industrialization began to infiltrate every aspect of life. Modernists were tremendously excited by the idea that the twentieth century was to be a gleaming age of progress in which engineering and technological advances would open up all kinds of new ways of living. In many respects it did and there are many today who fully agree with those early pioneers who wanted to see a halt to the aping of what they considered tired, historical styles and called for the abandonment of tradition. Henry Ford, the great über-industrialist of the first half of the twentieth century, was all for this. As he put it, 'We don't want tradition. We want to live in the present and the only history that is worth a tinker's damn is the history we make today.'

This is the context of Le Corbusier's famous definition of a house as 'a machine for living in'. Le Corbusier has become the figurehead of Modernism in the popular imagination. He has achieved that position because of his many ground-breaking, iconic pieces of architecture, but back in the early years of the twentieth century he wanted function to come before design. His dream was to see mass-produced housing made out of reinforced concrete with walls that were not always load-bearing so that the interior could be rearranged whichever way the householder wanted it. To clarify what he meant he produced a set of 'Purist rules' that refined and simplified his mode of design. It dispensed with ornamentation and pointed to a time when the production of buildings would become as efficient as a factory-floor assembly line.

Many dense, ugly, low-rise blocks of flats like this one were supposed to be part of a modern, transformed world, but rapidly became notorious ghettos of crime and deprivation. The open spaces that surrounded them often became the domain of drug dealers and remain a place for gangs not children's games, blighting the outlook and quality of life for many thousands of people.

From these principles came his 'Immeubles Villas', which were blueprints for a tower-block world. Half of central Paris would have been flattened to accommodate it. He drew up his plans in 1922 and they have often been blamed for the atrocities committed in the name of housing that shot into the skies above every European city in the 1950s and 1960s. I accept that this is not a fair accusation to make against Le Corbusier. He wanted to create an architecture that would lift people out of what were still in many cases dark, prison-like Victorian cities, but nonetheless Le Corbusier's ideas became part of the foundations of the homogenized conventional approach to many buildings today, which so often fail in my view, to enhance well-being, or destroy it altogether.

Le Corbusier's ideas became part of the foundations of the homogenized conventional approach to many buildings today, which so often fail in my view, to enhance well-being or destroy it altogether.

Those 1920s' plans for large blocks of apartments piled up in skyscrapers sixty storeys high were to have created a new metropolis on the North bank of the Seine and very much laid down the sorts of principles so many of our built environments still follow today. Le Corbusier could see, for example, that the car would become king, so he planned separate walkways for pedestrians that did not intrude upon the freedom of the motorized highways. He advocated a zonal approach whereby residential districts would sit separately from commercial quarters. And, at the centre of it all, he planned a huge transportation interchange for buses, trains and even aircraft. It was to have been an urban world of high-rise concrete blocks conceived on a matrix of straight avenues and highways.

Twenty-five years later, in the wake of the Second World War, city authorities throughout Europe seized on Le Corbusier's ideas and those of his fellow Modernists as a cheap way of solving chronic housing shortages. True, they were not Le Corbusier's designs. They had none of the integrated detail he had included in the tower block he did see built in the South of France. However, during the course of two decades, countless ugly, cheap and boiled-down versions of his 'vertical villages' were thrust up into the skies of Europe, most of them 'system-built', assembled from pre-fabricated panels of concrete that were lifted into position and held in place by bolts. These high-rises were arranged in 'estates' that often included a shopping centre also made out of slabs of grey concrete. This shopping hub was connected to the tower blocks by wide avenues that sometimes crossed wide green spaces. By the late 1960s, in the UK alone, 470,000 of these new flats had been built in this way, hardly any of them by architects, mostly by engineers. They represented the rather depressing fallout of Le Corbusier's high-minded and complex vision – in effect 'machines for living in' made entirely by machine – and all too quickly these 'modern' and 'new' towns lost their look of modernity and turned into hell.

I spent a great deal of time and effort during the 1980s and early 1990s trying to draw attention to the problems of such soulless, inner city estates. It concerned me greatly that so many people who had been accommodated in these convenient, concrete cul-de-sacs in the sky found that when their newness quickly faded and those wide green spaces became what they remain today, bleak and threatening no-man's lands, they were living in places that disconnected them from any sense of community, let alone Nature. Many of these estates became violent and soul-destroying ghettos with no sense of place nor any kind of beauty, and many countries still wrestle with the problems they create.

Even where there is more of a sense of community, I find that too often this same approach to design blights the townscape. Local schools and colleges, hospitals, supermarkets and shopping centres are too often built of the same breeze-block and steel, following homogenized designs and creating homogenized places and so go some way to debilitating the communal sense of well-being by abandoning the time-honoured, organic grammar of harmony.

Keeping the voters happy

To me, the way in which Modernism bulldozed away the complexity of the shape and form of organic processes is another example of mechanistic thinking at work. It served to encourage the comprehensive industrialization of life that now abounds. Even by the 1920s the consequent dominance of a Modernistic perspective at the expense of traditional ones, and with an urban psychology weakening traditional theology, the ground had became fertile for consumerism to seed and grow. The straight lines of Modernism, the desire for simplicity, uniformity, monoculturalism and convenience, all combined to frame the way millions of people in the developed world began to view life and the natural world, and so it remains. However, the view has not come about entirely by accident. All of the elements I have sought to explore in this chapter have been aided by some very sophisticated social and psychological manipulation.

As Tony Juniper, Ian Skelly, and I prepared this book, I found it revealing,

CRAVEN "A"

Will not affect your throat

not to say frightening, to learn the history of the advertising industry. It was very much the brainchild of Edward Bernays in the 1920s, who had been heavily involved in a wartime propaganda exercise to persuade a reluctant American population to support the war effort. He took these techniques and applied them for commercial and, indeed, political purposes, describing them as the 'engineering of consent'. Bernays realized that for the doctrine of consumption and unlimited economic growth to be successful, the desire to consume must never be fulfilled, but if people could be persuaded to seek that satisfaction, not only could it be a great engine of economic growth, it would also keep the population becalmed. Bernays's deliberate aim was to shape the desires of the American people as part of a new consumerism that was very much allied to 'keeping the voters happy'. He pioneered the use of the third-party authority of experts. Famously, he encouraged massive sales in bacon and eggs by conducting a survey of 5,000 doctors who said that eating a hearty breakfast is a good thing to do. In his book of 1928, which he seemed proud to call *Propaganda*, he says that 'the conscious and intelligent manipulation of the organized habits and opinions of the masses is an important element in

'We are governed, our minds moulded, our tastes formed by men we have never heard of … who understand the mental processes and social patterns of the masses. It is they who pull the wires that control the public mind.' Bernays defines his profession at a time when he changed the view of women smoking in public and boosted sales of cigarettes by paying models to smoke as they marched in women's rights parades.

democratic society'. No wonder, then, that we have an economic model predicated so firmly on the consumerism Bernays's techniques produced.

In no small way those mass-media techniques have maintained the only condition that makes it possible for mass consumerism to be sustained at its present high-octane level. That is, the blatant widespread disconnection of huge numbers of people from Nature's astonishing ingenuity and capacity for self-assembly. Mass consumerism can only remain successful if the advertising industry manages to convince us that the idea of living *with* the grain of Nature is a kind of cultural and technical immobility and that it is much better to pursue an ever more comprehensive kind of 'transgenic' manipulation that defies Nature, even if that means that at the very cutting edge of things we threaten to destabilize the very structure of life itself.

This was all part of the Modernists' dream – a machine-driven world that has now become computer-driven, where convenience is ours at the click of a switch or mouse. It is easy to sound like a curmudgeon when considering this aspect of our contemporary world but, given that at present children in many developed countries only spend half the time outdoors that they did twenty years ago, but an average of forty-two days a year online, should we not be more concerned about the long-term impact such a culture will have on the way our world will be when they grow up? On the horizon sits something called Caves, which are special rooms that create a 3D virtual reality. Research is also underway into the possibility of connecting the motor cortex of the human brain to the electronics of these games, so that the pulses from our brains that instruct our feet to move or our hands to wave can drive the virtual foot, or the virtual hand of an avatar in the same way. And then there is the so-called 'bionic eye', which is a contact lens imprinted with a tiny electronic circuit, and LEDs that would allow us access to the internet or to a virtual world as we walk down the street. It would be overlaid on top of the real world, creating a very strange hybrid world indeed but one, ultimately, that drives us ever closer to our machines and ever further from the living experience of connectedness that Thomas Aquinas saw slipping from our grasp all those centuries ago. This is certainly the concern of many educational psychologists who, from what I gather, say there is a very real possibility that the generations of the future will grow up having learned to ignore the essential living experience of existence. They argue that, contrary to what we hope it will do, our dependence upon technology could loosen our inner moorings or even sever them altogether. Some predict a generation who have only an 'instrumental' relationship with the world and the outlook of some sort of 'manchine'. Perhaps I am being alarmist here, but I very much doubt that a virtual world will be able to sustain

RIGHT: *The stylized view of traditional farming. But can we learn from how things used to be done when the approach was much more integrated? The horse not only pulled the plough, it fed the soil. More integrated approaches in modern farming could help achieve food security at the same time as promoting a range of environmental and social goals.*

us if we do not take more care of the real one. If we successfully eliminate all of the vital life-support systems from the real world, what use then an avatar? There will be no escape, virtual or otherwise.

Certainly these developments do not promise to nurture the sense of 'Mother Nature' that our forebears felt animating both them and the entire world. With three-quarters of the population of the United States now living in a city and the proportion much the same across much of Europe, her call becomes ever more distant. Many millions of us walk for most of the time on concrete paths amid forests of concrete and steel. We do most of our living and working housed in breeze-blocks and glass and breathing air in offices that is conditioned, not fresh. We buy our food from the supermarket, ready-washed, ready-chopped, ready-packed, all of it vacuum-sealed in plastic. Even the supermarkets are destined to be places of complete disconnection. It is quite possible even now to shop without ever speaking to anybody. Shopping online, of course, guarantees it.

Mechanistic thinking has forced more and more of daily life to cede to the corporate process where people are required to communicate with machines

A globalized economy has given rise to the mass use of call centre technology. This faceless monoculture has become a front end example of the dehumanizing impact on everyday life of globalization.

rather than with real people. The call-centre culture is fast becoming the norm, whereby customers are treated rather like a factory-farmed chicken, required to follow a specific route through a transaction that is designed to make the computer system operate efficiently. This may make economic sense – from a mechanistic point of view it surely represents perfection – but the price society pays is an erosion of its well-being. The increased subjugation to mechanical, monocultural procedures, laid down in premeditated sequences of questions and responses, avoids human contact and understanding, eliminates the need for human judgement or the deployment of common sense, the application of insight or the generosity of trust, the exercise of imagination or the sharing of compassion. In such a mechanical world, disconnection is complete and all human values are completely swept away, and I just wonder if we really want this to be the future ground plan of our civilization…

It was the American writer, Wendell Berry, who observed that industrialization and the impact of mechanistic thinking shift our notion of progress onto a straight road that promises we can somehow reach the horizon, as opposed to our true situation, which is part of the natural process of 'revolving in order to dwell'. It is a loaded shift from what he calls 'reproduction to production', which renders our approach to the world 'a failure on its way to being a catastrophe'.

I began this chapter by stressing that the journey our civilization has taken through history to get to this point is neither right nor wrong in itself. It is the course that history has taken; the milk has been spilt and there is no use crying over it. We now have to deal with the situation and make sure it does not happen again. To do that, we have to be clear what caused the problems that are now occurring – the deep causes as well as those that are more obvious – and I hope I have managed to give an outline of the deepest of them. What I have tried to describe is how the approach that has gone into making the significant discoveries the world has benefitted from has itself shaped the way we now tend to look at the world. It has trained us to approach the issues and problems we would like to overcome in a particular way because the history of our outlook has denied the crucial insight offered by that half of ourselves that has been excluded. I cannot help feeling that it would serve us well, as well as the rest of life on Earth, if we considered taking a step back from our present course and thinking again about the sort of world we want to live in – the sort of world we would want our children and their grandchildren to live in. As it stands they will accuse us of the very thing we seem so keen to aspire to, of being 'consumers'. What we fail to see is what they will see all too clearly, that in our voracious appetite to consume we have also been consuming their prosperity and undermining their well-being. I can hear them now, calling back to us through time, 'Why did you not *do* something?'

In that question lies the answer. They will ask 'why' whereas we have been concerned too much with 'how'. Empirical science and our tremendous strides in technology pursue the 'how' but, as I hope is so clear, it is not a discipline qualified to consider 'why do we do what we do?' We assume that it can, but the empirical science upon which our modern world depends cannot alone articulate and explore all the levels of reality.

We depend upon the world that depends upon us and to have any chance of surviving on this planet and sustaining both the biodiversity and the Earth's many 'eco services' that our survival hinges upon, we must begin to recognize the wider impact of our predominant way of thinking and begin to heal the disconnection it has brought about. Otherwise, no matter how fast and clever our computers become, we will go on initiating more and more chaos, more and more confusion and more and more dissonance. And with all that swirling around about our heads, we may never recover the balance of the essential, life-sustaining harmony that Nature needs for everything to endure.

Complexity within coherence, diversity with unity

Despite this nightmare scenario there is hope, and it comes from science as it does from elsewhere. Some of the most recent discoveries in science have revealed the reverse of what the mechanistic view of reality has always taught us to expect. In many fields, particularly in those areas studying microbiology and sub-atomic phenomena, what we now realize is what the ancients knew well, that there is a deep-seated interconnectivity present at every level of the physical world. It is for this reason that the physicist Werner Heisenberg, who gave his name to the Uncertainty Principle in quantum physics, would tell his students not to see the world as being made of matter. It was, he said, made of music. We have learned from Heisenberg that the physical world is not made up of individual parts but is essentially 'process and movement'. Particles 'dance' from order to disorder and back again, expressing in their dance a diverse set of movements that always happen within the defining boundaries of unity. Holding the very fabric of Nature together there is the pull of order and an integration that is balanced and harmonic.

It is also interesting that many scientists today now acknowledge that if the strength of the gravitational pull in the universe or the power of electrical repulsion were to be even slightly different, then the universe would be very different indeed and creatures like us would have little chance of surviving. This is known as the 'Anthropic Principle'. Some scientists hold it to be vitally important. It proves, they say, that the universe is not a random universe, but a particular universe. The scientist Sir John Polkinghorne has described it as 'alone capable of producing systems of complexity sufficient to sustain conscious life'. It has the freedom to be itself, he says, but is 'a value-laden world in which there is a supreme source of value whose nature is reflected in all that is held in being'.

David Bohm, who developed a theory of quantum physics that treats the totality of existence – both matter and consciousness – as an unbroken and ordered whole, said in 1994, shortly before his death, that 'in my scientific and philosophical work my main concern has been with understanding the nature of reality in general and of consciousness in particular as a coherent whole, which is never static or complete, but which is an unending process of movement and unfoldment'. The point here is that while we have undeniably made great gains with our many advances in science and technology since Galileo first pointed his new telescope so bravely at the Moon, the mechanistic framework that he and others helped to develop has, in more recent times, also allowed an unnatural arrogance to prevail in the gap left by the loss of this very precious understanding. It was once framed by Thomas Aquinas in theological terms and now it is being underlined in a contemporary way by the cutting edge of a science that is going beyond the mechanistic model and stressing a physics of relatedness. Unfortunately this has yet to trickle down into the way the world operates in the mainstream, where the disconnected, industrialized mindset still rejects as nonsense any notion that there is territory beyond the material, simply because it cannot be measured by empirical science. As it was once put to me rather prosaically, this dominant view may well have given us the smartest cars, but we have lost all sense of the soul. Our entire anchor, once embedded in the experience of being alive, has shifted to a mechanical, 'instrumental' relationship with the Earth and this puts us in a potentially precarious situation. We face high seas that I fear are going to grow ever more ferocious and we sail in a rather eye-catching, but flat-bottomed boat. There are many treacherous rocks and, rearing before us, the vast machine of economic globalization, which at the moment certainly does not wear a human face. It is quite clear from all of the evidence we now have that this leviathan needs to be tamed and given that human face, rather than the one it wears at the moment, that of the 'manchine' more at home in a science-fiction movie than ruling our real lives.

It is a fact of history that humanity will never wake up to the dangers until the crisis actually hits us between the eyes. Until then, vested interests stave off

the warnings by shooting the messenger or destroying the sanctuary that holds the wisdom of the heart, either by suffocating it with ugliness of all kinds or reducing the argument to the level of a tabloid travesty – anything rather than face the darkness that overwhelms the soul. At such a crucial moment in our history it is vital that we reignite the lamp and illuminate what has lain in the shadows.

It seems to me that this is the beginning of the answer, the root of the solution, which actually costs no money at all. It costs only human vision, imagination, ingenuity and human confidence at the ground level, in government and around the table in the corporate boardroom. If we hope to regain the balance – that is all; the balance – we need to restore in a contemporary way the best parts of the abandoned and ancient understanding of harmony that I have so far attempted to describe, and that includes a restoration of the spiritual perspective on life. There may be hundreds of thousands of flowers, a myriad array of trees, millions of animals and birds and countless billions of microscopic microbes, but each one testifies to the mystery of life itself. The trick would then be to take this restored perspective and to blend it with the best of our modern technology and our contemporary view of a globalized community to come up with solutions that work. This is what the next two chapters will explore.

There are many who are trying to do this around the world and it has been my good fortune to meet some of them and learn what they are successfully managing to achieve. I have also been at great pains to put this integrating idea to the test myself in all of the key areas that most concern me. In all cases what we will look at is what can be achieved if the philosophical outlook, the science and the technology are *all* in synchrony with Nature. From the creation of what are called 'virtuous circles' to much more integrative ways of teaching, the focus is on *including* rather than excluding our philosophical understanding of the wisdom to be found in Nature's processes and purpose.

LEFT: *Every ecosystem contains an interlinked diversity of life, where each animal and plant is dependent upon the health of its neighbours. Among the most diverse ecosystems on our modern Earth are the delicately balanced coral reefs.*

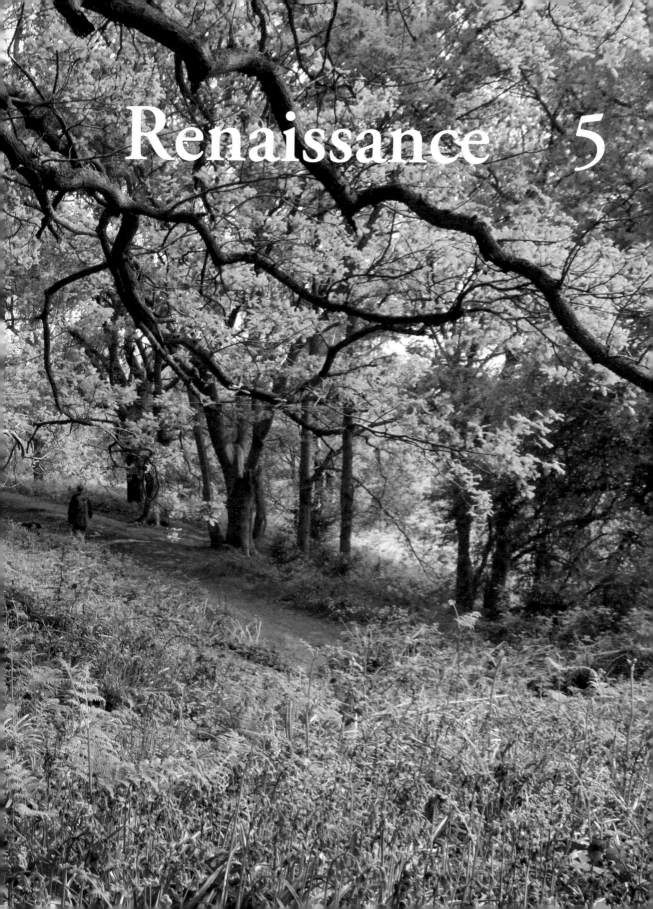

Renaissance 5

The new is in the old concealed,
the old is in the new revealed.

ST AUGUSTINE

Sometimes it takes an outsider to bring you up short. I am very fond of the story of Mahatma Gandhi, who did this in the 1930s during a visit to Britain. Asked what he thought of Western civilization, he replied, 'It would be a very good idea.' Although amusing, it was an acute observation about the direction that, even then, Western thinking was taking.

I am sure that the great emphasis in the industrialized world on material progress was not intended to destroy our planet's life-support systems, nor to cause humans stress and dissatisfaction, but I am afraid this is increasingly the result of how our civilization has developed. There is, as William Shakespeare put it, 'a tide in the affairs of men which, taken at the flood, leads on to fortune'. With ever-clearer reasons to be concerned as more and more evidence emerges of the combining problems we face, it seems to me that now is the right time to review our direction of travel and, as Shakespeare advises, to 'take the current when it serves or lose our ventures'. In other words, recalibrate the compass now or risk being lost for good.

For me, and many others, a way of living and looking at the world based on how Nature works and what might be called the 'principles of harmony' is a logical and attractive alternative to the way we operate at the moment. The core of these principles would be to approach things with a clearer sense of pro-portion and balance. It would require whole-istic, joined-up thinking; a deeper appreciation of natural beauty and the willingness to see the connections that hold the web of life together. We would need to value Nature's capacity to self-order her complexity; to recognize Nature as our guide, rather than seeing her as a machine that we can abuse to breaking point.

This may sound idealistic, but I say it with my eyes wide open and in full knowledge of the many problems that are now besetting the world caused by our not following this approach.

We have had, and are still being presented with, many disturbing wake-up calls. Recently, for instance, news emerged that for the fourth year in a row more than a third of honey bee colonies in the United States failed to survive the Winter, which means that more than three million colonies in the US and

PREVIOUS SPREAD: *Springtime in an English woodland. The timing of trees coming into leaf, the flowers appearing, insects emerging and the birds nesting are familiar aspects of a cycle that has been repeated for thousands of years.*

LEFT: *Mahatma Gandhi at No. 10 Downing Street in November 1931. Among the many great observations he made was that 'Happiness is when what you think, what you say, and what you do are in harmony.'*

billions of honey bees worldwide have died in recent years. This poses an alarming threat to food security around the world. The fact is that a third of everything we eat is dependent upon bees for pollination. Scientists say they are no nearer to knowing what is causing this catastrophic collapse, but there is plenty of evidence that modern pesticides have played their part. Given that bees, like nearly every other bug, are insects, I would have thought it was rather obvious, yet we carry on with this narrow-minded, mechanistic approach to industrialized farming with our focus only on high yields at whatever price, lacing the fields with pesticides that kill insects while being told that they do not affect bees. It is quite bizarre how we continue to put our faith in the very substances that are destroying the harmonic cycle which produces our food. It really is a form of hubris and I wonder if those who practise such irresponsible scepticism in these matters will ever see that the emperor is wearing no clothes.

I described at the start of this journey the scale of the crisis we are facing, in terms of climate change, the degradation of Nature and the depletion of her resources. I pointed out how our economic system was a big part of the problem and then set out how, as I see it, we are suffering not so much from an absence of technology or from a dearth of good ideas about policy. The real fault is that this way of thinking is utterly out of harmony with the rest of the world. I would like now to explore what could be the many solutions we might bring to bear on the main problems we face. Not only are there plenty of examples, but I intend to explain how they work and I would like to outline some of the thinking that many experts tell me needs to be adopted in the world of business and commerce, and at government-policy level, for these solutions to have a really long-lasting effect. This journey will take us through the various areas that I have concerned myself with during the course of this book. What we are about to do is explore different ways of farming that work with Nature rather than against her and see how viable they really are. Then, to investigate how sustainable building techniques and a more rooted approach to town-planning can conjure a more durable and liveable built environment that nurtures rather than depletes a sense of place and community. This is becoming ever more important in view of the huge population-driven expansion of currently unsustainable urbanization all around the world. I also intend to explore approaches to medicine that could be beneficial if they are allowed to be made complementary to mainstream treatments, and I want to give a taste of just some of the extraordinary innovations and designs that are now emerging from a brand-new form of engineering called biomimicry that is dedicated to following Nature.

Land and life

Perhaps the most direct daily connection that all of us have with Nature and its cycles is through what we eat. Nature feeds us, even though by the time much of our food is packaged and arranged under bright lights in a supermarket it is difficult to appreciate that fact. As we have seen, wrenched from its context, rendered anonymous, and displaced from its origin, much of what we eat is far from what can reasonably be described as natural. So how might we better approach agriculture so that we do not end up going backwards to an age of insecure food supplies, low yields or unaffordable diets, but instead rebuild our relationship with land and food and restore some of our connections with the natural systems that sustain us and all other life?

I am afraid there is still a widespread view of small-scale farm enterprises as quaint alternatives that appeal to niche markets, but there is increasingly firm, expert opinion that the future of farming must embrace methods that reduce the use of chemicals, rely less on monoculture and cut emissions of the greenhouse gases that add to climate change. However, working with Nature emphatically does not entail taking an 'unscientific' approach to food and

Supermarkets have brought us great convenience, but I wonder if at the same time we are losing something rather important, for example a meaningful relationship with our food. It seems to me that the more distant the origin of our food, the further we drift from Nature.

farming. It is very much a question of thinking through what science is for, and how we use our knowledge and understanding of the world we inhabit.

Soil and soul

While researching this book we came across Tim Waygood's farm in Hertfordshire in the East of England. Sitting on a shelf in his farmhouse kitchen there is a framed picture that shows him at his home at Church Farm in the village of Ardeley in 1969. In the photo he is two years old and walking through the middle of a battery hen house. He still lives at Church Farm today, now with Emma and their two children.

Tim's family farmed for generations, but in 1988 the 175 acres that had been in mixed use were finally taken out of production. Despite various steps towards more modern and industrialized methods, it became ever more difficult to earn a living from farming. The farmers best placed to make ends meet were those with bigger land holdings favoured by the official subsidies. Tim's parents effectively gave up. They decided to 'set aside' their land under new European Union rules that paid farmers not to farm and thereby to reduce the over-production that had resulted from earlier subsidies that offered an incentive to increase output. It was these official payouts that had, over the decades, led to the drainage of wetlands, the clearance of woods and the destruction of hedgerows. More chemical-intensive methods had helped cause precipitous declines in once common birds, animals and plants, all to produce massive surpluses of milk, grains and meat. Then came new financial incentives that were specifically designed to discourage farmers from doing any farming at all.

Tim gained a degree in agriculture, but didn't initially take up farming. Instead he set up several successful businesses, working with dozens of different companies, helping them to refine their strategies. As time went by he became more aware of what seemed to make for true success – which was good health, proper food, good relationships and a decent environment. He was also becoming aware of the different stresses that were appearing: climate change, diminishing resources, disappearing species and economic instability. This led him to begin farming in 2008, but not as his parents had done. He decided to turn Church Farm into an alternative to the global agri-business model with its system of supermarket outlets and the corporate-driven food-processing industry. And as a result his farm looks rather different from other farms in that part of North Hertfordshire.

The vision that Tim and Emma have embraced seeks to build a farm business around traditional methods in tandem with modern technology and knowledge.

Their strategy is to fight the economies of scale with what they call the economies of scope; that is, the advantages that come with diversity and local connections. They plan to put farming in harmony with Nature, and to put farming once more at the centre of the rural community. As a result, Church Farm is once more coming to life. Using traditional methods, some organic growing techniques and extensive free-range animal rearing, they aim to produce top-quality food for local markets using drastically less fossil fuel.

Woods and hedges are being replanted, grasslands restored and ponds reinstated. A walnut wood has been planted, and ducks and geese reared beneath the trees. Hundreds of chickens run free-range in orchards that contain more than 700 newly planted trees that will produce some 130 traditional varieties of fruit. Turkeys range in newly planted woods, foraging among the oak and ash saplings. Redpoll cattle graze on fields sown with pasture seed mixes from the late nineteenth century so as to boost their growth from what grows in the fields, rather than using imported animal feed. An old orchard that appears on maps from the early 1700s has the now rare Lop and Berkshire pigs foraging beneath the trees.

The areas of old woods that escaped the post-war ecological blitz are once more being managed traditionally to produce timber and fuel wood. The farm is also becoming a valuable social resource. The local authority use the farm as a day-care centre for people with special educational needs. One local person, who previously made the two-hour round trip to Cambridge every day for care services, now travels a matter of yards from his mother's cottage to be on the farm. As other schemes have found, for many people with special needs helping on the farm and being around animals is a far more beneficial environment than an indoor institutional setting.

Free-range chickens have become a modern symbol for consumers seeking a more harmonious and natural relationship with food. Animals can be reared in better conditions and this brings benefits for the environment and human health as well.

Tim and Emma are finding a ready market for their produce. They believe that many people are awakening to different ways of living and seeing, and that there is a renewed appetite to reforge connections with land and food. The battery hen houses are still there, but no longer as brutal tools of industrial farming. One was recently converted into a café and the other is now a farm shop. Both sell food grown on the farm.

The roots and inspiration of Tim's ambitions for Church Farm can be traced to his childhood experiences – for one thing he was always concerned about battery farming. He was also inspired by the many books he read and the various people he met who told him about different approaches to farming and food. I was delighted to hear that another source of inspiration came while he was at agricultural college in Reading during the 1980s. It was during this time that he heard about my own plans to begin organic farming.

A Duchy Original

The farm where I introduced organic methods more than twenty-five years ago lies on the other side of England, around Tetbury, on the Cotswold Hills in Gloucestershire. The Duchy of Cornwall Home Farm was, until the mid-1980s, typical of many farms in Southern England. For more than forty years land use here was shaped by official policies aimed at boosting production through industrial-scale monocultures. In common with what had occurred at Church Farm, hedges and stone walls had been ripped out, ancient pastures ploughed and yields were increased with artificial fertilizers and pesticides that were applied in liberal quantities, all as part of a new quest for 'cheap' food.

I found it a deeply depressing picture. Instead of natural fertility, the land was made productive with ammonium nitrate – a synthetic plant nutrient made using huge quantities of natural gas and water. The farm had also become isolated from the community and instead of maintaining local connections had become embedded in an ever more globalized food chain.

In the mid-1980s, after an intense debate and having faced considerable opposition, I finally managed to convert Duchy Home Farm to organic production methods; to step off the chemical treadmill and put the land into the agricultural equivalent of a detox regime – literally. It was quite a shift to move just over 1,000 acres of the farm into a more sustainable and localized food system. The decision was controversial, and I have to say that back then not many people could see the sense of my plan. But times have changed a bit.

Today Home Farm no longer relies on artificial chemicals; fertility is sustained by plants, animals and the careful management of everything by rotating the use of the land. Key to this approach is the natural nitrogen-fixing properties of clovers and the use of animals. For three out of seven years the land is planted with clover and grass. This is then grazed by animals, which return fertility to the soil in their droppings while at the same time they convert sunshine into meat and milk. The grass and clover is also cut to make silage to feed animals in Winter. To me it makes perfect, logical sense to farm in this way, in partnership with Nature, rather than in a destructive and exploitative relationship that, by definition, cannot last because it gives nothing back to Nature in return for what she gives us.

Organic systems effectively mimic the way Nature works. Like Nature, they do not produce waste nor pollute the atmosphere. The whole system is geared to the miraculous ingenuity of Nature and proves that this ingenuity can be harnessed not just to grow things, but to reduce the quantities of greenhouse gases in the atmosphere. The soil itself acts as a very effective carbon storage system, which means that organic farming techniques help us in the battle to mitigating climate change emissions.

It is vital that we pass on ideas about sustainable agriculture to younger generations. I am delighted that Prince William has an interest in farming. Here we are together walking around Duchy Home Farm back in 2004.

What is really important to understand is what organic farming is *not*. That is to say, it does *not* depend upon the use of chemical pesticides, fungicides and insecticides. It does not rely upon artificial fertilizers, the prophylactic use of antibiotics, any growth-promoters or GM technologies. Nor does it use

industrial rearing systems. Instead it recycles animal waste or composted organic waste to build up the soil's fertility; it employs homeopathic treatments for livestock wherever possible and those animals are fed on grass-based regimes, as Nature intended.

180 brown and white Ayrshire dairy cattle quietly chew the cud in the cowshed and here again the approach is different. The essence of the organic system is balance and so, to maintain the health of the livestock and of the land, the stocking rate is not too high. Antibiotics are also only used very sparingly, unlike cattle reared by intensive methods where such drugs are used to *prevent* disease rather than to treat it – an approach that has also progressively increased the prevalence of micro-organisms that are resistant to these antibiotics, including ones that cause disease in humans. The animals in the cowsheds at the Duchy Home Farm munch the silage collected from the fields in the previous Summer. It smells sweet and fresh. They wander around on a deep bed of straw which, with their dung, is periodically collected to make compost that, in turn, is used to increase soil fertility. With the arrival of Spring, the cows will graze outside.

Each field is periodically left to rest and put under clover and grass. When the animals graze this grass they add to the process of restoring nutrients naturally to the soil. Earthworm casts are everywhere. These and a whole host of other organisms work away at recycling organic materials, converting 'waste' into usable nutrients that will power and sustain the next cycle of growth. In this way it is natural processes that sustain productivity. There is no need for fertilizers made from fossil fuels.

When fertility is restored and the land is ploughed again, first wheat is planted. This crop needs high fertility; then comes a crop of oats, then beans and barley in combinations that bring down the fertility of the soil in a planned and optimal manner before, once again, the land is put back under clover and grass which is then grazed, and so the cycle of rebuilding fertility naturally begins again.

While the fertility of the land is being restored, the animals that grazed there provide meat and dairy products, thereby earning income and supplying food between rotations of arable production. The rotations are planned so as to disrupt the breeding cycle of pests and avoid the routine use of insecticides and herbicides. It is a harmonious cycle that works for both people and the land and which respects Nature's cyclical economy.

Vegetables are also grown – carrots, potatoes and onions, among others. These are mainly grown to supply local markets. There is a vegetable box scheme and a 'veg shed' where people can buy food directly from the farm.

Local schools and shops are supplied with organic produce as well. Grains and milk supply the Duchy Originals company that I set up in 1990 for making organic biscuits, bread and ales, amongst a range of other products.

It is not only the soil that has been restored. Since the organic conversion of Home Farm, some seventeen miles of hedges have been replanted. Using traditional skills, these living fences have been put back following decades of encouragement to have them removed. After watching these hedges grow I have greatly enjoyed learning and applying the traditional skill of hedge-laying. It is a surprisingly challenging and energetic occupation that tends to be pursued by refreshingly down-to-earth characters from all over England who wend their way to Highgrove for a friendly competition each Winter. The results are wonderful, for not only do traditional hedges create effective boundaries for livestock between fields, they also prevent soil erosion, help to harbour wildlife and shelter bird life and are just as beneficial to insects, which can otherwise be in short supply on intensive farms.

The farm is run as much as possible with closed loops or so-called 'virtuous circles', whereby the means to produce food is originated on site. For example, as was the case before the age of globalization, the farm woodlands are back in active use, with timber used for construction and fuel. Insects, birds and plants are benefitting from the resumption of traditional woodland management, in particular using a hazel coppice rotation that is a constant renewable, natural source of biomass fuel.

In addition to relying less on fossil energy and the emissions they cause, organic systems also help to protect the global environment in other ways. The quantity of organic material in the soil, for instance, is significantly increased so it stores more carbon and thus reduces the amount of carbon dioxide in the atmosphere. This, perhaps, should be more widely understood. One shocking estimate I saw suggests that the worldwide degradation of soils because of intensive industrialized farming systems has led to the loss to the atmosphere of between half and nearly three-quarters of the organic carbon in agricultural soil.

I would like to say that in light of the success of all these elements to create a successful organic farming system, there is wider cause for optimism, given the evidence that has now been collated by various bodies. The Rodale Institute in the United States, for instance, has calculated that if more sustainable farming methods were adopted to increase the organic carbon in the soil on the world's 3.5 billion acres of tillable land, then the farmland would be able to accommodate up to 40 per cent of our current carbon dioxide emissions. Moreover, their research estimates that this can be achieved with no decrease

Members of the pea family, including this red clover, naturally build up soil nitrogen in ways that are beneficial to other plants. By using rotations of such plants it is possible to increase fertility in soils to grow crops without resorting to artificial fertilizers. These increase emissions of greenhouse gases and are made from finite natural resources. They also cause serious pollution of the environment, leading to losses of biodiversity.

in farmer income or yields. In this way, organic and other, more sustainable farming systems become a method that converts carbon dioxide into a food-producing asset as it also improves the quality and structure of the soil – hence the principle of 'virtuous circles'. However, despite these sorts of findings, little will change until the resistance of the agribusiness Establishment to such findings has been overcome.

David Wilson has been the farm manager at the Duchy Home Farm for the past 25 years and he and I have also worked out a plan to begin experimenting with 'biodynamics'. This approach to food production takes farming a step closer still to the natural processes that govern agriculture and all other life cycles on Earth, especially in relation to the health and resilience of the soil – ultimately the most important resource we have. Early results have been promising. The farm is also experimenting with low-till ploughing techniques, which reduce the amount of diesel needed to power tractors and increase the organic matter in the soil – on both counts reducing carbon dioxide emissions.

I hope it is clear that the whole system is very much based on how Nature works, and thus is more in harmony with Nature than industrialized farming. There is minimal waste; the running of the farm is wherever possible done through cycles, loops and virtuous circles together with plant and animal diversity. Weeding is done mechanically rather than relying on herbicides and, I am pleased to say, the proof is in the pudding. We produce a great deal of healthy food.

And yet, despite the fact that I have just described what organic farming does *not* do and does *not* use, it is apparently not accepted in official circles that organic, or more naturally produced food products, are any healthier than the industrialized versions. I wonder why this is so when, as I mentioned earlier, such colossal sums are spent every year by British water companies to remove the pesticides and other chemicals that leach into the water supply. I can only presume that all that expense is necessary because such chemicals are considered damaging to people's health if they are left in the water supply? And what about the long-term effects of the overuse of antibiotics and growth-promoters in the farming system? Surely there is now sufficient evidence to show that the over-use of such things has had a profound effect on the human population's resistance to antibiotics? All these factors surely need to be taken into consideration when we seek to define the meaning of the term 'healthy food'.

A turning tide

I have certainly discovered the hard way that if an organic approach to the way we produce our food is going to work properly, all of the other dots in the picture must be joined up. At the moment that is not the case, certainly when it comes to what happens to the food once it leaves the farm gate.

It is remarkable how fast times can change and new ideas become the established way of operating. Today we have a globalized, 24/7 food supply system and this is now considered the norm in a large number of countries, and yet it is a very recent phenomenon. Before the age of cheap oil and mass transit systems using planes, vast container ships, motorway networks and huge distribution centres, most food was grown and sold much more locally.

In South West England a community group, called Transition Towns, interviewed older residents to gather stories of how life was in the town of Totnes in Devon before the age of oil-fuelled globalization. Some remembered how most of the vegetables were grown nearby, some in the middle of the town, including produce that was taken on barrows to shop fronts just a few yards away. Today those gardens have gone – replaced with supermarket car parks – but, I wonder, can we reverse this continuing trend by creating, for instance, local networks of farmers who supply local shops, schools, hospitals, prisons and defence establishments? It is not complicated. All it needs is good leadership and good organization.

Fortunately, some people are doing this. One example I have come across involves the catering manager at the Brompton Hospital in London sourcing food from a 'hub' of local farmers in Kent, opening up the hospital kitchens again, hiring a restaurant chef instead of having to rely on pre-cooked food brought in from elsewhere, reducing waste because patients were actually eating better food, cutting down on the constipation treatment needed to remedy bad diets (which costs the British National Health Service some £100 million a year) and, thus, achieving a virtuous and more sustainable circle. Ever since I saw this innovative and effective initiative in operation I have tried to interest other hospitals to adopt such an approach, but it is not an easy task.

Another idea already up and running that I have helped to promote works amid the rolling green hills and mountains of Mid Wales, one of the most beautiful landscapes in the British Isles. The land has been used for centuries for farming, especially for rearing sheep, which is about all that some of this landscape can sustain. Sheep farming, though, has become an increasingly marginal activity. Many farmers have given up trying to compete with cheap imports and, as a result, a source of local produce has declined. It is a victim of a narrow financial 'bottom-line' approach to agriculture which I feel very strongly needs to take a wider view.

Upland sheep-rearing is not just about food; it is an essential activity for maintaining the characteristic upland environment, as well as the human communities that depend on it. This is why in 2008 I helped to set up the Cambrian Mountains initiative with a group of farms in Mid Wales – to supply Cambrian Mountains lamb and beef while at the same time seeking to protect and enhance the incredibly beautiful environment there. The animals are predominantly Welsh native breeds and reared with traditional methods passed down in farming families over many generations. The lambs are born, grown and processed locally and produced in accordance with a set of environmental principles. It is, for me, another small example of harmony in practice.

The Cambrian Mountains initiative shows how sheep and cattle reared with more traditional methods could be an important part of a more sustainable rural economy, especially if people once more become used to eating older sheep in the form of mutton. Raising awareness about mutton, a once commonly eaten meat that has a really delicious flavour and texture, is the purpose of the Mutton Renaissance campaign. I initiated and helped launch this campaign in 2004 to support British sheep farmers who were struggling to sell their

older animals, and to encourage a renewed appreciation of the special quality of mutton.

I might add that I have also put a considerable amount of energy into highlighting another product from sheep-rearing that has also, incredibly, fallen into neglect – and that is wool. Some farmers now find so little financial value in wool that they are forced to destroy what they produce. On its own this is madness, but the picture is even more topsy-turvy than at first appears. It is an example of the ultimate insanity of our disconnected way of life. Owing to the supposed lack of demand for this truly renewable product, all because of the competition from man-made fibres that are based on oil products, a new breed of sheep has been 'engineered' that does not need shearing. Believe it or not, the breed is called 'Easycare' and there are already 500,000 of them not producing wool in the UK. All this, despite the fact that wool has so many beneficial properties and is endlessly renewable. Apart from its use in clothing, it offers a particularly effective form of insulation for houses; it is non-flammable, whereas man-made synthetic fibres catch fire much more easily; it does not produce toxic fumes and the kind of dust particles that can contribute

A shepherd and herding dog with sheep, Beddgelert, Snowdonia National Park, Wales. Traditional upland farming has created characteristic landscapes of great beauty, while at the same time sustaining livelihoods and producing wholesome food.

to the growing epidemic of allergies in the developed world; it is long-lasting and it has unique properties to retain heat. In short, it is one of Nature's gifts and we are either throwing it all away or breeding a sheep that does not produce it any more.

The reason I feel so passionately about wool and mutton is that they are products intimately linked to the maintenance of a continuous, harmonious pattern of existence and production that not only sustains many of the landscapes that help to define our identity and nurture our very souls, but also sustains entire communities of people who have a profound understanding of the local conditions. These people are the very backbone of what simply has to be a living countryside if we wish to sustain the future security of both food and Nature's services. This is why I have been trying to find ways of raising awareness of the potential for such a wonderful resource like wool and I am delighted that so many green building firms are now using it to insulate houses. However, it needs the volume house-builders to take wool much more seriously than they do if the tide is to be turned for good.

Variety of life

Likewise, we need farmers, growers and retailers to think about widening the variety of the produce they grow and offer for sale. I have already explained how Nature displays a tendency towards variety and away from uniformity and yet we seem to be heading in the opposite direction. Take something as simple as the humble apple. There was a time when thousands of different varieties of apples were grown in the UK. Some of these have been found to be nutritionally different from the modern varieties that are produced in globalized mono-cultures, not to mention their inherent resistance to many of the diseases that affect the modern hybrids. Some of the older apples are as much as sixty times richer in biphenols, an important nutritional element with antioxidant properties. Before apples were transported around the world on container ships using vast quantities of energy and producing huge emissions of greenhouse gases in the process, the UK enjoyed over 1,000 varieties. Different varieties also ripened at different times, ensuring a year-round supply. The indus-trialization of fruit production has meant the loss of this wonderful variety in favour of a few globally dominant types. What is also not a little depressing is that, amazingly, the majority of these are imported, even at the height of the English apple harvest. For me this is yet another symptom of our disconnection and short-sightedness that we are supposed to accept as the norm; a world that is effectively eliminating agricultural biodiversity in the name of 'efficiency'.

Farmers have been persuaded to rely on fewer and fewer varieties of almost every food species – usually those bred to do best in chemical and water-intensive monocultures. Some of us over the years have tried to draw attention to the long-term dangers of pursuing this course, but inevitably have been systematically shouted down.

Despite this, I am inspired by the various organizations and individuals who continue to fight against this erosion of diversity by running seed libraries and encouraging the propagation of the scarcer varieties. I know of many gardeners who are planting older varieties of vegetables, for instance, and then trading seeds with friends. Movements in developing countries are also fighting the swamping of local farming by genetically modified and other industrial-scale methods. And this struggle extends beyond the orchard. There are also increasingly active efforts to halt the loss of traditional farm animal breeds. Indeed, if it had not been for the heroic efforts in the UK of the Rare Breed Survival Trust over the past 40 years, we would have lost many more native breeds to extinction. I have been proud to be their Patron for very nearly thirty years and have done what I can on the farm at Highgrove to help preserve the

The rich biodiversity of fruit, vegetable and grain varieties generated by farmers through millennia of selective breeding is being rapidly depleted as we come to rely on the fewer and fewer varieties most suitable for globalized industrial farming methods.

Cattle at the Baylham Rare Breeds Farm, near Ipswich in Suffolk, England. This rare collection was threatened with destruction by the UK's first outbreak of bluetongue disease.

gene pool of a range of rare or endangered breeds. I have done the same with rare varieties of apples and vegetables.

The Survival Trust's work has focussed on the maintenance of endangered native breeds of farm animals, many of which were overlooked for widespread use during the 1950s and 1960s in the face of the industrialized, monocultured approach to farming. Many were even allowed to˙become extinct, or nearly so, because they were regarded as lower-yielding and thus of less immediate value. Yet many of these animals are better adapted to a range of different environments, including grazing conditions, and produce better quality meat because they mature more slowly. Thankfully, many traditional breeds have once more found favour, but they are in existence only because of the dogged determination and herculean efforts of a very few dedicated people who have struggled against the onslaught of the industrial leviathan.

These modest acts of resistance are helping to maintain the vast history of knowledge and vital biodiversity embedded in the different varieties of animals, vegetables, fruits and seeds that are still (just) with us. They are the result of the efforts of the long-dead gardeners and farmers who, over the centuries, honed traits in plants and animals that will doubtless be of immense value to us in the future – indeed the not-too-distant future. If we hope to have any chance of weathering the storms to come, then it seems critical to me that we maintain and use this diversity that has been handed down to us as the result of many hundreds of years of careful stewardship. The ball has been passed to us and we have to keep running in the same direction, *with* the grain of Nature, and thus towards a more harmonious food system, rather than sticking with the one that has taken root during the last half-century or so, however convenient it may have appeared to be. I might add that it is a system few people asked for or wanted. It is a system imposed from the top down on communities

and individuals whether they like it or not, with the justification resting on that somewhat nebulous idea of 'the market'.

Restoration

In addition to attempts to conserve crop plant diversity, various initiatives are under way to restore more natural ecosystems and to use these to produce food in more harmonious ways. For example, in the USA considerable effort is going into the conservation and restoration of North American prairie habitats, in part through encouragement for ranchers to employ a practice called rest-rotation grazing, in which some pastures are rested while others are temporarily grazed – a far more natural use of the prairies – rather than straining and twisting this fragile ecosystem to breaking point.

Others are going further by looking at the potential to raise bison instead of cattle. These great herbivores were driven to the brink of extinction during the nineteenth century, in part deliberately to undermine the economy of the

I am often struck by how so many people seem to have an almost instinctive wish to feel a connection with the land and cycles of growth. This is a garden and allotment at the St Pancras Almshouses in London. It seems that even in the heart of great cities many people want to reconnect with Nature's rhythms.

indigenous people. These animals could be brought back from their present low numbers as a more natural grazer, the value being in the fact that they are capable of producing meat without destroying the delicate prairie vegetation and soils. The North American native people understood this, and perhaps the more recent inhabitants of the prairies could learn much from those who lived there for millennia before the arrival of ploughs and machines. I must emphasize again that this is not to argue against the use of science and technology, but rather to combine technology with the wisdom that so often comes from closer coexistence with Nature. This wisdom is, itself, part of the 'science' of understanding and working with Nature, but it was arrogantly discarded in the twentieth century as being merely 'primitive' and irrelevant.

The modern 'conventional' view is that we must persist and expand the style of industrial farming that is characteristic of the North American prairies in order to 'feed the world'. And yet, there are plenty of people who are deeply concerned about genuinely 'sustainable' or durable farming and who believe that with the impact of climate change and the diminishing availability of 'cheap' oil, the prevailing view is, at best, debatable. It certainly seems to me that in the present circumstances we need a far more informed debate about the way to restore resilience to farm systems by mimicking more natural cycles.

Moves toward more sustainable farming on the North American prairies have led in some areas to the increases in native bird populations. This is a male sharp-tailed grouse in its dancing display during the breeding season on the Nebraska tallgrass prairie, USA.

HARMONY

After over thirty years of considering this issue and working on the solutions myself, I can say with some confidence that the only truly durable and resilient farming systems are those that are attuned with Nature. In other words, the ones that give back to Nature as much as they take from her. Only this kind of farming system will feed the world in the long term, and this is why I believe we need to begin right away with the transition towards this more natural form of farming.

I wish I could say that it is only a matter of time before this happens and that I am greatly encouraged by the changes that are now taking place. Certainly there are moves both in more affluent, developed countries, where some people are making strenuous efforts to adopt more sustainable approaches, and in developing countries where farmers are looking for ways to manage land with methods that lead to more secure incomes, to local food security and to create a more resilient environment. However, at the moment, any such moves are offset completely by the continuing pressure from the big retailers for ever more efficient production systems. Official thinking is just as bad. There are constant efforts to introduce GM technology and to encourage 'efficient,' economy of scale systems of farming which, of course, means that small farmers and entire communities are effectively driven off the land. Is this what people really want to see happen, given that genuine food security is intimately linked with communities of small farmers all around the world? It is important to recognize that if the kind of 'big' thinking I have mentioned is not tempered by a more rounded and integrated approach – and fast – then we will witness the comprehensive and disastrous destruction of both community and natural capital on a scale that is impossible to exaggerate. It is this big element in the game that has to be somehow persuaded to take a more rounded view.

Land, forests, and food

In 2009 I was lucky enough to see at first-hand the work of the Amazon Permaculture Institute, which has worked since 1997 to transform an area of deforested land near the Amazonian city of Manaus in Brazil. Not only had the rainforest been cleared, but the soil had become degraded and, as a result, farming was rendered impossible. The group of organizations that established the Institute, including the National Council of Rubber Tappers, set out with the aim of showing how it is possible to produce food without destroying the rainforests.

I was staggered to hear how the United Nations Development Programme estimates that in Brazil alone there are some 65 million hectares of degraded

I have been enormously impressed by different small-scale projects that show how it is possible for us to work more with the grain of Nature. In this picture Carlos Miller (left) and Ali Sharif explain to me efforts to save the Amazon rainforests in Brazil. During this trip I was very flattered to be declared a 'Friend of the Forest' by the Governor of Brazil's Amazonas state for speaking out to save threatened ecosystems.

land. Placing this vast area under much more sustainable and biodiverse eco-system-based systems of production, combined with rainforest restoration, could not only produce huge quantities of food, but could also take pressure off the rainforests and so preserve the vital services these ecosystems provide for the world. This is the thinking behind the work of the Amazon Permaculture Institute.

Their ambition, I discovered, is to create a fully integrated sustainable system. And this is indeed what they have done. Water is collected from rainfall and used water is cleaned and recycled on site. Vehicles are powered by used cooking oil collected locally while a methane biodigester uses pig waste to make the gas that powers the cooker in a café. Although it works on a small scale, the aim is to demonstrate to both rural and urban communities how it is possible to engage in genuinely sustainable farming practices – again working in harmony with Nature rather than against her.

Today the land restored by the permaculture project is covered with trees that were selected to fix nutrients and restore the chemical balance in the soil. There are pigs, cows, chickens, goats and fish ponds. I was told how more than 5,000 students and farmers have so far visited the project, taking away with them a powerful inspiration as to what is possible when a different view of the land is adopted by the people who use it. The project provides a stark contrast to the ravages wrought on the Amazon basin by the globalized market for commodities, especially for timber, beef and soya beans, produced with exploitative and destructive industrial methods.

The message from the project is that with careful planning and the correct approach it is possible to produce food and livelihoods from formerly degraded and abandoned land.

I came away from seeing their work convinced that it would be possible to create a much more diverse tapestry of agro-forestry in many parts of the world. Such an approach would help to ensure biodiversity 'corridors' that could link existing forest and plantations that are not just vast monocultures of one type of tree or crop. It might also be possible to empower local community associations to manage this pattern of biodiverse agri-forestry so that they could better care for their own environment. Not only would this enhance 'Nature's capital', it would enrich what I would call 'community capital'.

Once you start looking at what is happening you realize just how many projects like this there are in the world. There are many examples in the tropics alone of how people are trying to establish a much more harmonious relationship with the land and with the natural systems that support people's livelihoods. One example is the uplands of Sri Lanka where a landscape that was once entirely clothed by dense tropical rainforests has been cleared to make way for tea plantations. Unfortunately, these farms are now encroaching on what is left of the forest – what is supposed to be a World Heritage Site. And so a local organization has begun a programme to encourage the farmers to adopt a similar approach to agro-forestry. They aim to reduce the use of agro-chemicals, to intercrop tea fields with other plants and to put in place measures to reduce soil erosion. Hundreds of farmers have taken part in training courses and have learned how to improve soil fertility using the shade of trees and hedgerows that also provide fuel wood and fence posts. The foliage is used to make green manure.

The Luangwa Valley in Zambia is typical of much of sub-Saharan Africa: an undulating landscape of mixed woodland dotted with smallholder farms. The main crops produced for local food are maize and sorghum. Increasingly, however, and in part because of official policies to promote the growing of crops for international markets, farming has shifted from growing local produce to cash crops like tobacco and cotton. The effect of this has been to reduce soil fertility and to increase soil degradation. Having been forced to cultivate new land, farmers are encroaching on forests – including protected reserves that

harbour animals including eland, hartebeest, kudu and elephants. Local trees have been cut to provide the wood fuel used to cure tobacco.

As farmers make this shift they are more inclined to poach wild game and to make charcoal from trees to augment their diets and incomes. To improve this situation, in 2002 something called Community Markets for Conservation set about teaching more sustainable farming methods to stabilize the loss of the forest and to conserve wildlife. Farmers learn about zero-tillage farming, how to apply home-made fertilizer individually to each plant and how to use the previous year's crop waste rather than burning it. As a result the soil holds its moisture, weeds are suppressed and the soil has become more fertile. Poultry has been promoted as an alternative to game meat and honey production has been encouraged to provide an additional source of income. Farmers also receive assistance in processing, packaging and marketing their produce, and within a very few years these farmers' food security has been improved, nutrition has diversified and wildlife populations are more stable.

In China, too, there are hugely inspiring examples of how even grossly degraded land can be brought back into productive farming, in the process slowing down soil erosion, reducing the build-up of sediment in rivers and so lessening the risks of flooding. The Loess Plateau in North Central China, for example, is an area the size of Belgium which, over the course of ten years, has been completely transformed. What was a barren, brown landscape a decade ago, a place completely denuded and degraded, has been brought back to life in the most miraculous way. The comprehensive programme began with the planting of trees, particularly on hilltops. These trees made sure that the soil could once again retain the moisture when it rains, rather than all the water flowing away down the hillsides, leaving nothing but desert-like conditions behind. Once the structure of the soil was restored, terraced fields were built on the degraded hillsides and the soil in them was augmented with plenty of organic matter. These terraces were planted and managed using a careful system of crop rotation.

The transformation is staggering. Large areas of once barren and unproductive land have returned to life, producing food and sustaining entire communities. I was utterly amazed to see the difference, and hugely encouraged because this same approach could be applied to many other similarly blighted wastelands, providing an ecosystem-based approach to agriculture which, in addition, can help to mitigate the effects of climate change.

I have come across many other examples of programmes and projects that are trying to introduce more benign approaches. I have followed the progress of a French organization called Action Carbone, for example, which manages a wide variety of schemes. They have distributed solar cookers to communities in Bolivia and Peru to replace those that run on wood, gas kerosene or bosta (manure), thereby replacing cookers that emit up to a tonne of CO_2 annually with ones that run off the Sun and do not produce harmful fumes. In Madagascar they have launched a scheme that turns the organic waste of the city of Mahajanga into compost for local agriculture, reducing by 120000 tonnes the amount of CO_2 emitted during the running time of the project. In India they are deploying small-scale anaerobic digesters that recycle kitchen waste and animal manure. These improve the living conditions of farmers and their families and offer the possibility of protecting the Indian forests from over coppicing. They also have a programme running in Niger which plants forest and fruit trees to improve food security and decrease poverty in a population badly affected by the reoccurring food crises there.

I could go on, but I hope it is clear from this snapshot of the kinds of transformation which can be achieved that for genuinely sustainable food production to be achieved in many developing countries, the main problem is not an absence of sophisticated technology. Much more important is to engage the power of that 'community capital' I have already referred to. Empowering local people and helping them to improve soil fertility with local resources is far more productive. Education is also an important factor to get right – an education that teaches connections and certainly puts an emphasis on using water more efficiently and sustainably. It is worth noting here that common to so many of the success stories I have just mentioned is that the farmers were encouraged to grow older, traditional crop varieties rather than modern hybrids that do not need such profligate quantities of water as their modern counterparts, the hybrid varieties. And last, but not least, the whole system needs to be underpinned by good microfinance – of the kind that does not trap people into debt, but supports them.

New priorities

The focus in all these cases is much the same: to move away from policies and practices that promote what is often the myth of 'cheap' food and towards those that promote resilience and durability. In this way, livelihoods become more secure too. Local production reduces the dependence upon fossil fuels, which cuts pollution. Reducing the use of chemicals also cuts pollution, so people's health stands a better chance, as it does for animals and for what is, surely, our ultimate resource and the essential basis of a secure and common future – the soil.

I am not saying that all food should be produced on a small scale and for local consumption, or that international trade in farm produce is automatically undesirable. And I am absolutely not saying that we do not need technology. We do need it, but it should be part of a more balanced and comprehensive system that works as benignly as possible with Nature rather than through a vicious circle system and a technology that only solves one problem without looking at what caused it in the first place, never mind the other problems that might be caused in the process of solving the one in hand.

Considering the enormous challenges we face to preserve vital biodiversity, to prevent soil erosion, restore tree cover, harvest and conserve water, store carbon in soils and trees, revive degraded land and produce much-needed food, it is surely incumbent upon the large agri-business companies to say how their future operating model will be fit for purpose? At the moment I simply do not see this being articulated, but instead hear only attacks on the shortcomings of what are still regarded as alternative systems.

Curiously, and in direct response to some of the critics, what I am advocating here in many respects amounts to a more intensive approach to agriculture that produces higher yields. Recent research has shown that in Brazil, for example, they have enjoyed increases of up to 250 per cent in the yields of maize by using green manures and cover crops. Meanwhile, on Nepalese farms there have been yield increases of 175 per cent where similar agro-ecological techniques were adopted. In Tigray, in Ethiopia, the location of the terrible famine that shocked the world in the mid-1980s, yields have increased by a factor of three to five

since people began composting instead of using chemical sprays. These are all encouraging results and I am very pleased to say they have found backing from one very unexpected quarter.

This came in 2008 when a comprehensive review on the future of farming was published. At a packed press conference in London 400 experts, drawn from across the world who had been working through a process called the International Assessment of Agricultural Knowledge, Science and Technology for Development (IAASTD), set out a vision for the future of food production. I expected this exhaustive process to have concluded that all the answers lay in yet more technology and industrialization of production, with plaudits for genetically modified crops, more sophisticated chemicals and the need for even more extreme monocultures. But it didn't. Instead the report concluded that continuing with these approaches would exhaust our resources and put our children's future in jeopardy. It said that a narrow focus on increasing food production would undermine our agricultural capital and leave us with an increasingly degraded and divided planet.

The authors called for a different way forward for farming, and for new organizations and different economic and legal frameworks to combine

Ethiopian farm at sunrise. Infamous for a terrible drought and famine, some farms in this East African country are emerging as examples of resilient and sustainable agriculture. Bucking trends of recent decades, farmers have improved food security without the need for artificial chemicals. It seems that in many circumstances traditional methods can produce sufficient food, if the conditions are created for them to thrive.

productivity with the protection and conservation of soils, water, forests and the biodiversity of life in all these natural resources. Crucially, they concluded that the way to help the poorest farmers was not through the sale of expensive seeds, pesticides and fertilizers, but by helping them to revive or adopt traditional methods suited to their land – to render them less vulnerable to the failure of a single monoculture crop or a sudden drop in the price of global commodities which are factors that are always utterly outside the farmer's control. No matter how hard they work nor how many chemicals they have bought, these external factors, of course, only worsen their crippling debts.

This all proves to me that a durable system has to be resilient to such outside factors, and that means the roots have to be strong. In other words, for a community to enjoy strong food security, it must rely on a system that works from the roots up.

More natural and nutritionally balanced diets can improve educational performance. From field to plate, there are many reasons why we should wish to have a more harmonious relationship with our food. In this picture pupils tuck into a wholesome lunch at St Martin in the Fields School, Lambeth, South London.

It must be locally based and self-sustaining and not dependent upon outside inputs and the vagaries of what comes down from the top – just like Nature, in fact, where everything grows from the roots up, not the other way round.

Sustenance

At the other end of the food-making process I think there are many more solutions worth considering. One thing I find deeply troubling is how, for many in modern societies, there is a great chasm in their understanding of where food comes from and how it is produced. Many schoolchildren in more developed countries, especially those in families on low incomes in urban areas, rarely see fresh vegetables in the kitchen, never mind watch them growing in the ground. For these young people the connections with food, and thus the natural cycles of production, are truly severed.

They are cut off, literally, by the packaged and processed nutrition that lies between them and the land, the source of the sustenance that feeds their minds and bodies. Not only is there commonsense reason to be concerned about this, but also plenty of research. Some shows how children fed with properly prepared organic food for school meals behave better and concentrate more than those fed with fast food that is high in fat, sugar and salt. In some schools, parents have taken the initiative and run schemes themselves to provide high-quality, wholesome meals for children.

Food for Life, run by the UK Soil Association, is a brilliant attempt to reunite people with a more sustainable food system. It provides schools with specified proportions of organic, unprocessed, local food and builds direct relationships between farmers and consumers by operating school visits to farms. Teachers have told me that

they have seen noticeable improvements in the attentiveness of children who, for the first time, are eating properly while at school, after a reduction in the consumption of highly processed foods. There has also been a dramatic reduction in food miles and a cut in food waste, although the latter issue is one to which, in my view, we need to pay much more attention. In fact, I am amazed at how little attention we pay to this particular scandal. Between the farm gate and the dinner plates of many better-off consumers, a vast quantity of food is lost and destroyed. In some Western countries about a third of the total food produced is thrown away, which means that about 30 per cent of the land used to supply food to some of the richer countries is actually feeding bins rather than people. In the UK this waste costs about £10 billion a year and it should make any sane person question the claims that the world suffers from a food shortage. Clearly, the principal criticism of organic farming and other

There is growing evidence to show that an approach to food security which harnesses the best of modern science alongside traditional methods could be a more sustainable way to fight hunger and improve nutrition.

low-input methods – that it produces less food and therefore needs more land – is at best partial.

And yet, some critics argue that with a fast-rising population it is better, if not necessary, to farm land intensively, thus reducing the need to clear more forest and other natural habitat to increase the farmland needed to match rising demand. But this is far too simplistic an analysis.

For one thing there are now as many obese people as hungry ones, which suggests that the challenge is more about the types and patterns of nutrition available and how food and the means to buy it are distributed, rather than simply a question of the quantity of food produced. I have, anyway, seen plenty of evidence to show how in many cases small-scale organic methods can produce more food per hectare than industrial farming, while also supporting the livelihoods of more people.

I am aware of a substantial and growing body of evidence to show how, up to a certain point, the world's increasing population can be fed most successfully in the long term by agricultural systems that manage the land within environmental limits, restore and enhance ecosystem services and maintain soil fertility through the use of crop rotations and the recycling of organic wastes, all of which minimizes the use of non-renewable inputs.

By looking at the food system as a whole, from field to kitchen, and then considering the many questions of human and environmental well-being that are raised, an holistic approach is certainly far more logical and desirable than the fragmented model of industrialized farming that has dominated in recent decades. This is quite apart from the fact that in many places it is perfectly possible to produce more food using low-input and organic methods.

Cheap globalized food, bereft of identity and produced at massive environmental cost, holds huge risks for humankind, at many different levels. A more harmonious relationship with land and food – and thus ultimately with Nature – can deliver improved health and food security for people if we embrace the more integrated and holistic approaches that can take us there. I believe that if we allow Nature to be our inspiration, rather than be slaves to industrial-scale technology, then we could get there rather quickly.

I feel that we are entering a period in which many people are beginning to realize that Nature's limits are not shortcomings that need to be 'fixed' but guidelines we need to understand and work within. So, perhaps we are on the brink of major change. I certainly hope so, because we most definitely need to be. If we continue as we are, it will not just be Nature that is a casualty of the short-sighted industrialization of agriculture, it will be humanity too.

There are signs of a resurgent interest in reconnecting with food. From the popularity of farmers' markets to the huge demand for allotments, and from booming seed sales to people keeping chickens in the garden, there is good reason to believe that a quiet but profound cultural change is taking place – a new kind of globalization – one of ideas for sustainable living – occurring from the bottom up. This is in stark contrast to the globalization of the monoculture, which has been forced on us from the top down.

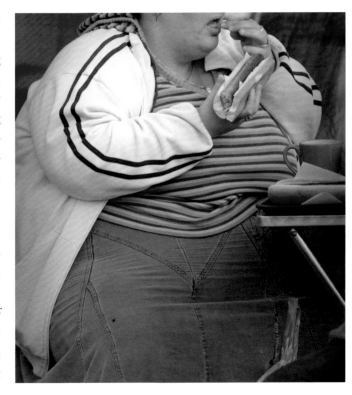

As we saw earlier, our welfare is at risk because of things like rapid climate change, depletion of natural resources and an accelerating loss of natural diversity. I hope I have managed to show how all of these things can be addressed in large part through a more durable, natural, holistic and harmonious food system. But there is another sphere of our lives where there are opportunities to improve both our own well-being and that of Nature, and that is the way we look at health.

Obesity is a major threat to public health. Since both malnutrition and over-eating are associated with low incomes it is perhaps too simple for us to focus on 'cheap food' as the solution to the world's nutritional problems. A more integrated approach based on a fuller appreciation of nutrition, health and environmental challenges is required.

Well-being

For an organism to be healthy it must be in harmony. The converse is that a body is 'dis-eased' – it does not enjoy an equilibrium. So, although we cannot see it, our health depends upon harmony and that extends to the impact of those external things that influence and shape our experience of and responses to the world. In a moment I shall explore one of the biggest of those things – what the jargon calls 'the built environment', which means our homes, our towns and our cities. I believe very strongly that a sense of place matters: an atmosphere can be beneficial to or destructive of our well-being. First, though, I want to concentrate on the influences on our health and how our modern systems approach healthcare and medicine.

One obvious public-health challenge that confronts many societies is obesity. Perhaps the most worrying aspect is the dramatic change in the shape and average weight of many children. Some argue that this is the greatest health challenge of all and I can well believe it. In England alone obesity among children aged 6 has doubled in the past ten years, and it has tripled among 15-year-olds. More than a fifth of adults are obese and about 40 per cent are overweight. Being too big leads to a whole range of health problems like type-2 diabetes, high blood pressure and abnormal fat ratios in the blood, all of which can increase the risk of heart and artery disease and strokes.

This drift towards obesity did not occur by accident: there are many forces at play, and a lot of these are not in the realm of what would normally be regarded as health policy. One is the very wide availability of highly processed, high-fat, high-sugar, high-salt, carbohydrate-rich foods, an avalanche of unhealthy eating that is in part driven by the appetite for fast foods and snacks. Promoted with hundreds of millions of pounds' worth of advertising, these kinds of foods have become completely embedded in our culture. They are widely available and cheap at the point of purchase, as a result of a wide array of economic and policy factors ranging from farm subsidies to weak controls on the messages used in promotion and marketing; the list is long and varied. But what would the impact be if we did turn to more organic farming methods of food production?

They produce less meat, of course, but many health experts suggest that we should consider eating less meat for the good of our health. The meat that does come from organic methods is of a better quality, so it is healthier. Organic dairy products also have different and healthier fat ratios, so they are also perhaps better suited to the most recent health advice.

Clearly, lifestyles have a big part to play in our health too. It stands to reason that if children are driven to school, rather than walking or cycling, then they are not going to be as fit as they could be. There are many reasons why this happens. Some relate to a parent's outlook, of course, but others are circumstantial and they beg the question how might we convert our car-dependent systems and lifestyles into ones that make it easier and more attractive to walk and cycle? That can at least be encouraged by planning decisions which, at the moment, all too often still place priority on cars and locate shops, leisure facilities and other daily needs miles away from where people live. Of course, as I say, it also comes down to an individual's choice in many matters. Far be it from me to suggest that we should all watch less television or not sit in front of computer screens for so long every day, but we have to recognize, surely, that if we encourage our children to enjoy these sedentary forms of leisure to the exclusion of all others then they are unlikely to be as fit and as healthy in later life as they could be.

The lesson that can be learnt from looking closely at the causes and

treatments of this modern epidemic, and others too, like stress and psycho-logical problems, is that good health actually depends upon more than well-funded health services and the availability of sophisticated drugs. Far more critical are those less obvious environmental factors that maintain the prevailing culture. Let me explain.

Communities for health

Many people and communities are, of course, taking action to curb these problems, calling for safe cycle routes to school, changing the food available when children get to school, and so on. Planning decisions often lag behind these calls, with the emphasis still on policies and measures that will promote narrowly defined economic growth and so-called 'competitiveness', rather than healthy and sustainable communities.

We came across a wonderful example of this in the beautiful county of Cornwall, which is undoubtedly one of the UK's most popular holiday destinations. It is also England's poorest county, where serious poverty and deprivation occur within some of the most spectacular landscapes in England. One place where social deprivation has been a serious problem is the seaside town of Falmouth, where the Beacon and Old Hill housing estates are the

Falmouth in Cornwall, England, is one of many towns and cities that face huge social challenges as economic circumstances have changed. The experience of a project here showed the potential benefits that can be derived from deliberate attempts to empower people through building what might be called 'community capital'.

poorest in the county. Falmouth once possessed a thriving port that employed thousands of people. Before the Second World War, housing estates were built on the steep-sided hills above the docks for the workers. The grey houses may still remain, but the jobs in the docks have more or less gone.

Over a period of years, the area became known for unemployment, poverty, drug abuse and crime. As a result, many families suffered from a combination of health and social problems. The situation became more and more desperate, until two health workers decided to do something about it. Their approach was not to distribute more medicine, nor to chastise those apparently responsible for worsening health statistics, for crime and the general spiral of decline that the community had become trapped within. Instead, they set about empowering residents to take their own steps to build strength into what had become a shattered community.

They arranged meetings for people to air their views about what had gone wrong and, crucially, to allow them to propose what could be done to improve matters. For the first time a wide range of official agencies were encouraged to take a more integrated approach and residents were given the chance to become organized. They formed a residents' association to enable discussion, to take decisions and to give the community a stronger collective voice. Thus began a process that led away from community anger and frustration and towards the establishment of a positive community spirit.

People raised money for community resources, which included the conversion of empty shops into community centres. The relationship between them and the official agencies improved and this led to the more effective delivery of public services. Houses were upgraded so that they were no longer cold and damp and were fit for families to live in. Insulation meant that houses were made more energy-efficient, and so more environmentally benign. Trees were planted and lighting improved, which changed the feel of the place. Residents of the estate began to tend their gardens and to plant flowers. Some of the drab grey houses were transformed with brightly painted cladding that not only saved heat, but also brought the neighbourhood to life.

The effect of all this was remarkable. I am told that within a few years the overall crime rate was cut by half, unemployment was cut by over 70 per cent, children's educational achievement doubled, post-natal depression was cut by nearly three-quarters and the number of unwanted teenage pregnancies also fell. There was a 50 per cent cut in child accident rates and a drop in asthma and chest complaints. The improved well-being was not only dramatic, but was achieved in a very cost-effective manner – all by using the strength and power of what had been a latent community capital.

Human beings are among the most complex of all life forms, and yet it seems to me that we sometimes regard our collective and individual well-being as something equivalent to looking after a car. We mend the parts as they fail rather than seeking out and securing the causes of health, which tend to include wholesome food, rest, relaxation, exercise, a sense of community, enhanced by the quality of surroundings, relationships and contact with natural spaces.

An environment for well-being

A few years ago I was very interested to note the findings of a report from the UK's Royal College of Physicians that set out how our modern epidemic of allergies is, in part, caused by our increased exposure to a variety of chemicals and, more remarkably, our lack of exposure to farm animals. One source of this experience is, of course, farming. So, surely, for this reason alone it would be better to focus more on sustainable farming methods, in that these have an indirect effect on improving public health. This is partly what I mean by looking beyond technological fixes and seeing things in a far more integrated way. It is all interconnected. An emphasis on industrialized farming methods, producing huge amounts of cheap meat and high fat, high carbohydrate foods results in a rise in the number of cases of type-2 diabetes, heart attacks, strokes, and several kinds of cancer. An emphasis on a less intensive approach, together with better, more integrated town planning and so on, can reduce these figures

While we have seen huge strides in medical treatments in recent years we have heard rather less about the therapeutic benefits of being closer to Nature. A wide range of studies now demonstrate that Nature is good for us, even the simple step of allowing patients to see green leaves and trees from a window can aid recovery from surgery.

and achieve a whole lot more besides, not least jobs in new technologies. This could all come with a lower-carbon society, especially if we planned change so that we could all have more contact with Nature.

Recent research has confirmed what I have suspected to be true for a long time – that we rely on contact with Nature for our general well-being. The official agency charged with the conservation of Nature in England – Natural England – has assembled plenty of evidence that people enjoy better health if they have access to green space and natural areas. Seeing and being in green areas and among trees reduces blood pressure, the heart rate drops and even our brainwaves change to relaxed alpha waves. What is more, the researchers suggest that people who have access to green spaces tend to live longer. Intriguingly, this effect is particularly noticeable among the poorest in society.

I am not at all surprised that academic studies have found that contact with Nature can make people more resilient to illness. For those who are ill, this contact certainly helps to keep their condition more stable. One well-known piece of research investigated the recovery rates of patients who had access to a view of trees in a Texas hospital courtyard, compared with those whose window gave sight of only a concrete wall. All the patients had undergone the same gall-bladder surgery so were considered to be in a broadly comparable group, except for the environments they lived in during recovery. The findings were remarkable.

Those patients who enjoyed a view of trees out of the window spent fewer days in hospital, they used fewer narcotic drugs, had fewer complications and registered fewer complaints with the nurses responsible for their care. Subsequent research built on the original theme to find that even Nature-related imagery, such as paintings and photographs on the walls, reduced blood pressure, cut anxiety levels and reduced pain. Research found the reverse to be true as well. Patients who could not see images of Nature suffered increased depression, were in need of more pain relief and spent more time in recovery. It seems that being exposed to the patterns produced by Nature are directly necessary for our health. That is quite a finding, and surely should have major implications for how we plan healthy societies.

Even the most mainstream hospital designers now acknowledge that Nature imagery is a simple and effective way to improve health, but it would be a tremendous advance if architects really took this evidence to heart by putting the patient at the centre of the design process and before the technology of a modern-day ward so that the experience of being in hospital provides contact with the natural world as well as medical care. At the moment, such is the frenzy for Health and Safety that hospitals in the main are Nature-free zones, and even flowers from visitors are now banned in many wards. How can we have reached such an insane situation?

I touch on these findings to underline how harmony and connection with Nature is not some vague or fringe concern. The truth is that there are real benefits for people, if only we can find the practical ways to respond to what is increasingly apparent from both common sense and science.

Complements and alternatives

The fact is that Nature is our best guide. The way our bodies work closely reflects how all systems in Nature work. As I mentioned earlier, rivers flow just as our blood flows, by the virtue of spirals. The Earth cleans itself using the same phenomenon – weather patterns are based on spirals. As we shall see a little later in this chapter, enhancing this principle has turned an observation into a brilliant piece of technology that massively reduces the amount of energy needed to shift water. Everything is dictated by the spiral motion. Just as water spirals down the plughole, so the fibres of our bones knit together in a spiral motion, as do the muscles of our hearts. And each of our organs keeps us alive and healthy on the basis of predictable cycles.

This is to say that our digestive process, our body temperature, our kidney and liver activity, lungs, heart and gall bladder all work to related rhythms, making our bodies, like Nature, self-regulating. The pH, or acidity/alkalinity, of our blood and cholesterol levels remains remarkably constant, for instance. In common with natural systems, our bodies are also adaptable. They change

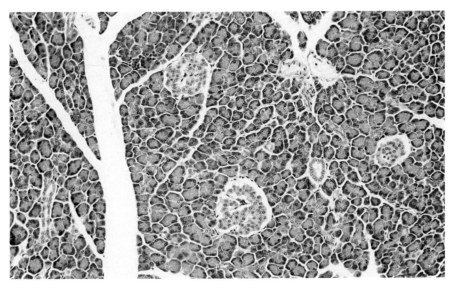

Our bodies remain in balance through self-sustaining systems. This is the human pancreas, which among other things regulates blood sugar. The way bodies maintain equilibrium mirrors the way that Nature works. Better health can be promoted through understanding the self-regulating systems that are at the heart of how we thrive.

colour in the sunshine, they can recover from jet lag and can become accustomed to higher altitudes. We also have the most amazing powers of self-repair if we cut ourselves or suffer bruises or fractures.

All of this reflects the way Nature works, and it was these characteristics, and in particular the powers of self-healing, that were once at the heart of medical philosophy. Hippocrates, the figure often recognized as the father of Western medicine, pioneered an approach based on the systematic restoration of balance to the body's equilibrium by enabling the body to benefit from its natural powers of recuperation and recovery. Eating well, exercise, massage and relaxation were at the heart of these treatments and today we know from modern science that he was very much on the right track.

The organs of the 'organ-ization' must operate in a connected way for our health to be maintained. So, if this is the case at all levels in Nature, why is it, I wonder, that we find it so hard to see this when it comes to looking at

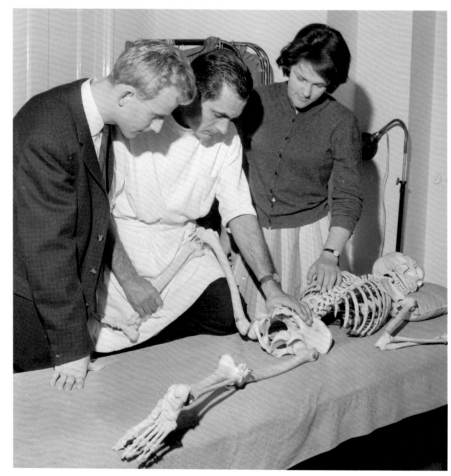

Students at the British School of Osteopathy in Buckingham Gate, London learn the principles of osteopathic technique on a skeleton. This approach to health rests on the fact that different parts of our body interact. Many people benefit from osteopathy, even though it is still often regarded as an 'alternative' to mainstream methods.

ourselves? Again, it seems our outlook is predominantly mechanical, and brings us back to the shortcomings of our mechanistic view of the world. So perhaps it is time for us to reflect more on the wider causes of health, rather than to rely so comprehensively on pharmaceutical remedies for sickness.

One approach to health that reflects this philosophy is osteopathy. A medical doctor in the USA, Dr Andrew Taylor Still, founded it, and opened his first school in Kirksville, Missouri, in 1892. Osteopathy was based on his conviction that the body's healing capacity can be facilitated by working on the inter-relationships between its structures, function and motion. Taylor Still called it osteopathy 'because you begin with the bones'. He was emphatic that disease was a disturbance in the natural flow of blood; the nerves cause muscles to contract, which compresses the blood returning to the heart. The bones were the starting point, the levers, used to relieve pressure on blood vessels and nerves. It is a fascinating method of treatment.

Osteopathy recognizes that ill health occurs when the ability of the body to adapt to a situation is disrupted. For example, an old ankle injury (which would be deemed the cause of the present problem) may have strained and altered the mechanics of the foot, destabilizing it (which is the effect). Over time the way in which the ankle joint alters to bear the weight (the adaptation) affects the alignment of the back and the muscles in the neck and the way the person walks. So if a patient complains about pain in the lower back and headaches (the most recent effects), these things can be indirectly related to the old ankle injury.

Osteopaths have a skilled, specialized sense of touch that can detect an altered quality of motion in the tissues of the body. Recognizing the state of disharmony in the body, they work to restore balance towards a state of harmony and health. In this way, osteopathy offers an integrated and preventative system of healthcare. It understands the unique connection between the physical, mental and spiritual make-up of each person. Rather than the prescription of pharmaceutical preparations, treatments focus on the diet, lifestyle and environmental and genetic factors that contribute to each person's health. Many osteopaths also work with what have become known as complementary and alternative healthcare methods.

Make no mistake, I am the first to celebrate the achievements of modern medicine and the many benefits it gives us and the way it immeasurably improves human well-being, but I still believe that there is a great deal to be gained from complementary treatments. Alas, even something as practical as osteopathy or acupuncture still sits on the sidelines under the heading 'alternative', inviting suspicion among the public and arousing a kind of angry

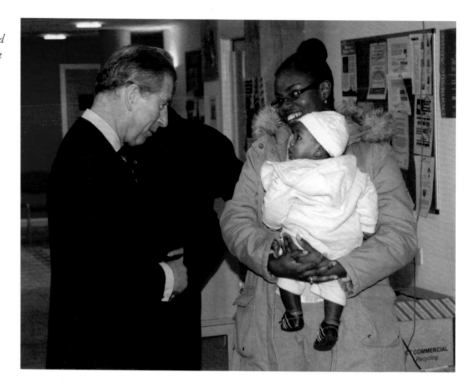

A wide range of conditions can be eased with medical care that remains on the fringes of mainstream medicine. We need to recognise the power of modern and traditional medicine working together. In this picture I am on a visit to Wellspring Health Living Centre in Bristol, England, where I presented an award for Good Practice.

derision from the mainstream medical profession. I have been pursuing this issue for nearly thirty years because I cannot bear to see people suffer unnecessarily when, so often, a complementary treatment can be beneficial, particularly for those whose condition has not been improved by conventional treatment or who have chronic disorders which often respond better to complementary therapy. It is a very frustrating business. For instance, one of the big arguments used against homeopathy is that it does not really work medically. The criticism is that people simply believe they are going to feel better and so they *think* they are better. They have responded to the so-called 'placebo effect'. It is for this reason that critics of homeopathy argue that it is a trick of the mind and its remedies are nothing more than sugar pills. What none of those who take this view ever seems to acknowledge is that these remedies also work on animals, which are surely unlikely to be influenced by the placebo effect. I certainly remember that when I started to introduce homeopathic remedies on the Duchy Home Farm, farm staff who had no view either way reported that the health of an animal that had been treated had improved so I wonder what it is that prevents the medical profession from even considering the evidence that now exists of trials of homeopathic treatments carried out on animals? It is not the quackery they claim it to be. Or if it is, then I have some very clever cows in my shed!

Suffice it to say, an integrated approach, combining the best of modern and traditional treatments, should sit comfortably with modern health challenges. Some 80 per cent of the UK's National Health Service time and resources are dedicated to chronic conditions such as allergies, heart disease and back pain. Chronic diseases are often treated with the same medicines used to treat acute conditions, yet in some cases it can be helpful to seek 'systemic' change in order to deal with long-term imbalances. This is what many so-called alternative treatments seek to do.

I am told that many of the functions that control balance, self-regulation, and restoration are located in the more primitive part of our brain and neural structures that are beyond our conscious control. Some so-called alternative treatments seek to work with these functions to aid recovery, relying on our innate capacity for self-repair, and this is the basis for several forms of so-called alternative therapies, including osteopathy as I have just explained.

I have long felt that healthcare budgets could make more use of complementary and alternative approaches. Whenever they are made available it is interesting that there is always a good take-up. For example, the clinic at Poundbury, the urban extension to Dorchester in Dorset which we will come to shortly, provides cutting-edge, modern gynaecological treatments, but also offers complements in the form of a professional homeopath, a hypnotherapist, an acupuncturist and a nutritionist. Patients find that these services, offered alongside more mainstream modern treatments, are true complements to conventional medicine and enjoy popular support.

In another part of the UK, in Northern Ireland, I have been impressed by how the government there has taken a lead in demonstrating the potential for complementary and alternative treatments. Following persistent encouragement from my Foundation for Integrated Health, a modest one-year pilot scheme took place. This offered patients, as part of their official NHS treatments, therapies such as acupuncture, aromatherapy, homeopathy and massage. The initial focus was on musculoskeletal problems, depression, stress and anxiety, and the study found that integrating complementary and alternative approaches with conventional medicine can indeed bring measurable benefits to patients' health.

Some 700 patients took part in the study and 65 per cent of them reported significant health improvements. Half of the GPs involved said they had reduced the amount of prescribed medication while the same proportion reported that their patients needed less frequent referral to hospitals. Four-fifths of patients said their general well-being had improved, while over half said they had been able to reduce the use of pain-relief drugs.

Overall the results showed how both patients and the health service can benefit from more integrated approaches to healthcare – a finding endorsed by the fact that all of the GPs said they wanted to continue to offer integrated healthcare.

This, to me, is another example of how a joined-up approach, based on principles of harmony, can help us do better for people and cut excessive costs. Just think for a moment how it could reduce the overall annual drugs bill. In the UK, it stands at more than £8.2 billion and it is growing at a faster rate than our Gross National Product.

He was emphatic that disease was a disturbance in the natural flow of blood; the nerves cause muscles to contract, which compresses the blood returning to the heart.

I have also learned from leading experts how we can understand a great deal about the causes of ill health through more traditional methods of diagnosis – for example, through examination of the iris, ears, tongue, feet and pulse, very much the basis of the Indian Ayurvedic system. This is not to say that modern diagnostic techniques do not have a vital role, but let us not forget what we can gain by using the knowledge and wisdom accumulated over thousands of years by practitioners who did not have access to today's technology. In fact, an over-reliance on technology can often mean that the subtle signs of imbalance revealed by an examination of the eyes, pulse and tongue are totally missed. Including the fruits of such knowledge, gleaned over 8000 years of studying the relationship of the human body to the rest of Nature and to the Universe, can but only provide an extra, valuable resource to doctors as they seek to make a full diagnosis. Why persist in denying the immense value of such accumulated wisdom when it can tell us so much about the *whole* person – mind, body and spirit? Employing the best of the ancient and modern in a truly integrated way is another example of harmony and balance at work.

It is, I think, not so surprising that traditional methods can deliver huge benefits when you consider how they are the fruit of such vast experience. The experience and learning could be of immense value to us in the modern world.

If combined with modern understanding then together they can achieve far more than either can alone.

As we face the challenges of the twenty-first century, with all its economic, social and ecological stresses, it seems to me more important than ever that we make sure we carry every effective tool in the box. To see one modern approach as the only effective and reliable replacement for all the others is most unwise because, as I have shown, it could leave us increasingly exposed in a more uncertain and disrupted world.

This is just as much the case in another critical area of human endeavour, one which affects human health and that of our planet, and that is the design of the built environments we construct and inhabit.

Towns, plans, and buildings

Over the past 100 years or so urbanization has undoubtedly emerged as the single most important trend shaping social conditions, and it is set to be more important still. Indeed, we have just entered what history might come to regard as humankind's first truly urban century.

The pace of change is utterly remarkable. In 1900 less than a sixth of the world's 1.5 billion people dwelt in cities. Since then the proportion of people living in rural areas has been in decline and we reached an historic point in 2007 when a United Nations report declared that, for the first time ever, more than half of the world's population lived in towns and cities – that is, half the 6.6 billion people who then lived on Earth. Projections for the twenty-first century suggest that by 2050 between two-thirds and three-quarters of us will be living in built-up environments and, by then, the world will most likely be home to about nine billion people.

Since the late eighteenth century the trend towards urbanization has run in parallel with industrialization. As societies have industrialized, so they have created more and more urban areas. This was unavoidable. People are always going to seek economic advantages where the best opportunities exist and that is always going to be in those centralized locations where businesses and economic transactions are most concentrated.

As the world economy has become more globally integrated, so cities have become more global in character, losing some of the qualities that previously distinguished them from each other. Such are the pressures on them that many urban areas have become increasingly degraded and unmanageable, hindered by either outdated planning systems or a lack of any planning at all. Those that have been managed have tended to be the victims of more centralized strategic spatial planning, driven by national governments' attempts to remain globally competitive. This has certainly helped to maximize their economic performance, but the trouble is that so many of them tend to look the same and feel the same. In this age of top-down globalization we have lost sight of the timeless and ecological approaches that could be of such benefit to our urban communities. So I would like to explore some of the solutions I have been trying to suggest over the last quarter of a century.

While cities and towns are both a cause and an effect of economic growth, it is clear that they must increasingly accommodate a lot more functions than simply providing venues for business. In many respects recent attempts to enhance global competitiveness have been to the detriment of what urban areas need to do for their inhabitants.

When I began looking at these sorts of problems over 25 years ago I could see very quickly that this trend towards sameness was driven by a combination of competitiveness, a monocultural ideology and a top-down globalization. I began trying to remind people that, until very recently, a town reflected its locality. This sense of local identity was created using local materials and traditional methods. Architecture and the building arts, until recently at least, were often based on the principles I touched on earlier – those traditional approaches that can be traced back to the ancient Egyptians and were passed down, fuelling the Renaissance and certain movements beyond that time.

The transmission of traditional knowledge in architecture and the related craft of building carried with it, as I have described in Chapter 3, an under-standing of how Nature fits together. This perspective, though, fell into decline in the modern era so that by the twentieth century approaches to architecture became ever more deconstructed, mechanical and divorced from the human and natural proportions that previously had informed building design. What often went with it was a sense of place and the traditional aim of reflecting beauty and creating harmony, aims that gave soul to the built environment. Modernist design rejected Nature as its guide and embraced the machine instead, as I described in the last chapter, and that approach led to the creation of soul-destroying environments that continue to be the source of a great deal of social blight.

My long-held belief is that the knowledge and tradition of the pre-modern period in architecture is of far more than academic interest or merely a matter of taste. Given the circumstances we face now, the rediscovery of more traditional approaches to urban development offer us some potentially valuable solutions, if only we would take the time to look at what is possible. Cutting emissions, saving resources and building places that people want to live in that

Aerial view of a housing development, Boraceia, Sao Paulo state, Brazil. Too many new developments seem to be designed more for cars than people. As a global peak in oil production looms, settlements set out like this are inherently vulnerable.

OVERLEAF:
A view of the Earth at night reveals where electricity use is most concentrated. Note the bright light emitting from the main industrial economies of Western Europe, North America and Japan. With the exception of South Africa the so-called Dark Continent remains aptly named.

will last are all potential benefits of adopting a more traditional or timeless approach, particularly when they are married to the best and most appropriate forms of modern technology.

I want to go further, however. Beyond architecture, design and technology, perhaps the most important resource in any built environment is the knowledge, relationships, values and perspectives held in communities. Although it is probably stating the obvious to say that people who live in a particular neighbourhood know best what they need, what their priorities are and how best to reflect these in decisions about where they live, I have seen how the modern trend has been towards more and more central planning. Communities get what others decide for them – there are no alternatives on offer. A top-down approach to planning has been something of a partner of the industrial-scale, copybook urban scheme. It is driven by the brutal economics of 'growth' and competitiveness and the pursuit of efficiency targets that care little as to whether a place ends up with 'soul'.

Yet how to work with communities to shape decisions is a far more important consideration than is often appreciated. As they found in a deprived community like Falmouth, if people are empowered to work together, there are huge benefits. This is why I have put such an enormous effort into developing something called Enquiry by Design, which is now widely used by my Foundation for the Built Environment. This method of decision-making sets out to use the knowledge, judgement and intuition that exist in communities themselves so that they shape the future of their built environments – whether in relation to regeneration or to new building.

I believe the ability of people to self-organize can be a remarkably powerful force, but sadly it is an opportunity that is too often untapped. Centralized spatial planning devised by specialist planners, trained in a twentieth century mechanistic ideology, sometimes misses fundamental choices and can lead communities in directions that are not in their best interests. I have enough experience now to know for sure that if people had been put more at the heart of the planning process, some of the disastrous urban environments created in many cities during the twentieth century might easily have been avoided. They would not now be causing the problems they are and, in some cases, would still be standing and flourishing. As it is, many are now in a state of decay whereas others became 'old' so quickly they were bulldozed to make way for the same mistakes to be made again.

I have seen how the power and potential of self-organization can be so effective in some of the world's poorest communities. Mumbai is one of India's largest cities and the country's financial centre. This great metropolis is

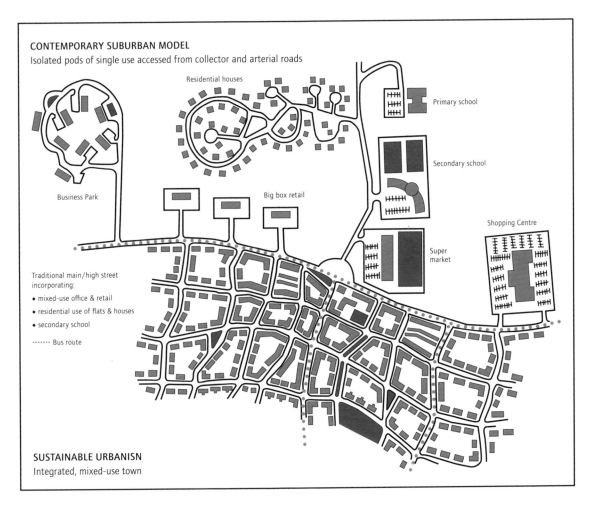

CONTEMPORARY SUBURBAN MODEL
Isolated pods of single use accessed from collector and arterial roads

Residential houses

Primary school

Secondary school

Business Park

Big box retail

Shopping Centre

Traditional main/high street
incorporating:
• mixed-use office & retail
• residential use of flats & houses
• secondary school

•••••••• Bus route

Super
market

SUSTAINABLE URBANISN
Integrated, mixed-use town

increasingly associated with the glass and steel skyscrapers familiar in many financial centres across the world, but it also has Asia's largest slum – an area called Dharavi, where 600,000 people are crammed into less than a square mile. I went to Dharavi and heard from people living there how the area is the result of earlier attempts to remove poor, homeless people from other parts of Mumbai. It grew until it was packed full of dwellings built by the occupants. There was no central planning, but it is amazing to see how people became organized nonetheless. Property rights and boundaries are understood, so are the rules concerning the sparse communal facilities – such as the few toilets. In many countries there continues to be an intense discussion about the rights and wrongs of municipal waste collection and how best to achieve recycling targets, but the people of Dharavi manage to separate all their waste at home and it gets recycled without any official collection facilities at all. It is not done in safe conditions and few people would want to do this work – but that is not my

The contrast between the typical suburban model (top) that wastes space and promotes discreet zones of land use, rather than a more integrated design (bottom) which conserves space and aims to create cohesive communities. The approach at the top of the diagram is the one all too often handed down by professional planners.

It may look chaotic, but urban design more under the control of the people who will live in the settlements they create often leads to a building of social capital and community resilience. This image shows the slum dwellings of Dharavi in the foreground with the Bandra Kurla Complex of Mumbai rising in the background.

point. The real lesson I took from Dharavi was about the vast asset we can call 'community capital'.

Some 70 per cent of the people who live in Dharavi also work there. Out-migration is low, and even people with the means to leave stay there. The slum has built up its own financial sector, with community banking enterprises using the savings of residents to extend loans to borrowers. This works on the basis of personal relationships and the power of the community to ensure the creditworthiness of those who borrow (somewhat in contrast to the recently imploded financial sector in the West). How such a level of sophistication and order can emerge from what outwardly appears to be chaos is an important question that surely holds lessons for modern town planning.

It strikes me that the tendency of most planners to make decisions for people, rather than with them, is one of the main reasons why we see so many disastrous results. Decision-making that places more emphasis on hard structures than it does on the knowledge and ability that communities have to decide for themselves misses the main point about how communities work.

I have found that many observers of communities like Dharavi quickly note the absence of physical assets such as power, water and sanitation, rather than the presence of this immensely important, but less tangible element of community capital that enables many apparently poor societies to organize themselves from the bottom up. The temptation is also to see their material poverty, rather than the rich complexity and diversity that holds the community together, despite such trying and uncomfortable circumstances – to see deprivation rather than how order has naturally emerged through people's complex interactions. This achievement is what particularly struck me about Dharavi. When you enter what looks from the outside like an immense mound of plastic and rubbish, you immediately come upon an intricate network of streets with miniature shops, houses and workshops, each one made out of any materials that come to hand. I am sure, having seen this elsewhere, that such a layout comes from a kind of intuitive patterning of the sort I explored in Chapter 3 – when I described how sacred geometry works and how it is inherent in humankind. It mimics exactly the patterns found in all traditional settlements. It embodies the ways in which human beings naturally create their communities. It struck me there and then that, throughout the twentieth century, Western ideology has disrupted the natural patterns of our humanity

Until recently the patterns of settlements emerged from meeting practical human-scale needs. This village in Malawi is laid out according to traditional local practice. It looks similar in layout to comparable scale settlements in other parts of the world.

and consequently created fragmented, deconstructed housing 'estates' or ghettoes that ignore or go against the grain of our intuitive senses, thereby at once introducing dissonance and dis-ease.

I am not saying that self-organized slums, lacking in basic facilities, are a model for future urbanization – far from it. What I do suggest, however, is that

I was very pleased to visit Jamaica during a tour of the Caribbean in 2008. Here my wife and I are planting a tree in Rose Town, Kingston. My Foundation for the Built Environment has been developing a programme of urban regeneration here which uses traditional building methods as part of a wider effort to restore a sense of community cohesion.

we have a great deal to learn about how complex systems can self-organize to create a harmonious whole. There are rules that exist to enable people to live using the wisdom and relationships they are born with, rather than relying on the devised vision of planners who might have little idea what actually works best for people on the ground. It seems that communities work best when the principles that Nature depends upon in order that she is self-sustaining are applied to urban planning and architecture. These principles bring together patterns of complexity and a sense of overall unity that, in turn, creates an order and a harmony that reflects Nature's own patterns of self-organization. Like Nature, communities thrive from the roots up, not the other way around.

It was with this perspective in mind that I encouraged my Foundation for the Built Environment to become involved in the regeneration of Rose Town, one of the most depressed parts of Kingston, the capital of Jamaica. I visited the area in 2000 and was struck by how the decline of the built environment and community had run hand in hand. In my Foundation's approach, the restoration of the social infrastructure has been as much to the fore as the design of buildings. Of vital importance has been the way in which my Foundation has worked with the grain of local people's culture and ways of life in order to reflect those realities in the design process. Ironically, it emerged that previous attempts to provide purpose-built housing for local people had failed because nobody wanted to live in it as it signally failed to reflect the cultural realities of their existence. Young people have been encouraged to equip themselves with the skills needed not only to restore and build the houses and other buildings but, even more importantly, the fabric of community. The result is that the houses are being restored and constructed with local materials and the work is being done by the people who live there. It is another example of community-building from the bottom up, and in a direction that leads to an enhanced sense of place and distinctiveness. And it is working as well – perhaps most tellingly in the steady decline of the gang violence that used to cripple the area.

It seems to me that where we live is very much more than simply a collection of buildings that keep us warm and dry. We need far more than that, and what we need is often best delivered at people level. A community's ability to organize itself and the flow of information that results can lead to living spaces that are very different from those envisaged by the agencies who sit at arm's length from the communities they plan for. Planners' and engineers' priorities are often more aligned with how best to achieve a narrowly defined set of objectives that conform to a now outmoded twentieth-century outlook rather than how to enhance and optimize the quality of human relationships emerging from the designs they hand down.

The vast expansion of urban areas that will occur over the next few decades will generate many environmental and social challenges – on top of those already with us. We now have to think very seriously about creating places (not housing estates) that people want to live in, but also places that can accommodate the various ecological crunches that are now close by. Cities will require more power, water and construction materials and so they will become an even greater strain on the Earth's already stressed life-support systems and capital resources. And if food has to be grown more locally, where will this happen if all is concrete and steel? I have been working to discover if there is a way of meeting these sorts of challenges.

Towns fit for the twenty-first century

Since the early part of the twentieth century, and particularly since the Second World War, the cities of developed countries have become increasingly surrounded by suburban sprawl. Urban areas have become progressively more segmented and zoned, with a separation between the places where people live, work and shop. This is partly why social divisions occur. If you create these various zones, better-off areas will become cut off from poorer ones and you end up with ghettoes, with a lot of space wasted and a high dependence upon the car.

As I see it, these tendencies have created several rather important vulnerabilities. If we build towns that need cars for every aspect of life, we create an excessive dependence upon endless supplies of cheap oil. By segmenting cities and their suburbs, putting different kinds of development in their own zones, cars can get about them, but not people on their own. So we set in place a long-term dependence upon something which is a finite resource; one that, in the near future, is expected by many experts to be in shorter supply. Zonal development risks the loss of social cohesion. If different levels of society live separately there is less chance of a cohesive, wider community and that can allow for social tensions to become overpowering. This does not happen so easily in traditional urban developments and, given that these challenges are not going to go away, might it not be worth looking at how these traditional approaches work?

This is what I tried to do when the Duchy of Cornwall, a 660-year-old estate, was confronted in the late 1980s with new house-building targets – particularly for families on lower incomes. This was a decision by Dorset County Council and it concerned land owned by the Duchy on the edge of Dorchester. There was no choice in whether a development should happen, the question I raised was, what kind of development should it be – one that was a slavish repeat of previous approaches to house-building; yet another soulless housing estate designed primarily around cars that wasted acres of precious space; or one that tried a new approach? The result was Poundbury, a new urban area contiguous with the ancient town of Dorchester.

By sitting down and consulting with local people through Enquiry by Design, the innovative approach to community planning that I have already mentioned, we were able to ensure the support of local stakeholders. Securing the support of the professional Establishment was another and far tougher matter but, despite that formidable hurdle, building finally started in the early 1990s and today around 1,800 people live in Poundbury which, although connected to an existing town that dates back to Roman times, is not a suburb. It has been designed as a series of urban quarters with a centre in its own right. Construction continues, but the town paradoxically does not feel particularly

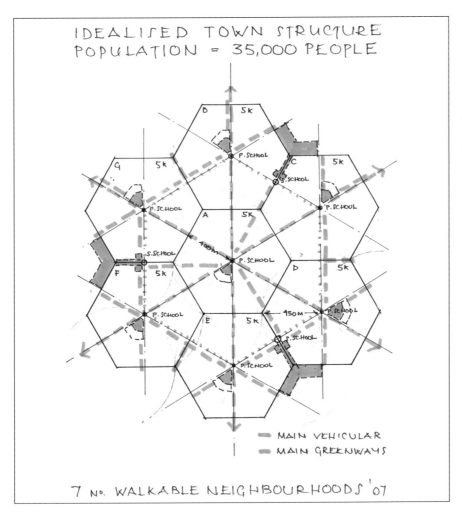

It is perhaps not surprising that traditional town structures based more on how far people would walk are more nurturing of the community that is so vital for the quality of the places we live. Here is the sort of structure we used to plan Poundbury where the entire Masterplan was based upon placing the pedestrian not the car at the centre of the design. Rather than houses slotted into a matrix of straight roads, the walkable town is created around a series of crossroads, mirroring the way traditional villages once developed.

new. Squares and streets are designed to have timeless, more traditional proportions and are given a vernacular identity by the careful use of local materials. And, as a result, its residents tell me that the place feels as though it has a soul and a heart.

Poundbury comprises high-density, mixed-use buildings; workshops, offices, local services, private housing and social housing – even factories – are all placed next to each other. There are no zones. The private and social housing sit alongside each other, together with workshops and other businesses. They are all built to the same standard, using the same high-quality materials. The aim is to create diversity and avoid monoculture. We devised a masterplan that, in turn, was related to a code from which the architects who designed houses and other buildings had to work. This masterplan was one that the community itself had helped to create and it set out a town plan that was inspired by approaches to urban layout that pre-dated cheap oil and cars. One of the main principles was that Poundbury should be a 'walkable town'. That is to say, most daily

Poundbury is an attempt to create a fully liveable settlement based on traditional approaches, and to do this in a way that truly meets modern challenges, for example reducing reliance on dwindling supplies of oil and building the social capital that is so important for our collective well-being.

needs should be within a ten-minute walk. This is the way nearly all villages and market towns are found to have developed. They would often begin by being a settlement on a crossroads. Over time the settlement would expand, but it is fascinating that, in so many cases – often until the advent of the motor car – the expansion of many settlements was always limited by the distance people were naturally prepared to walk. It is one reason, perhaps, why London is really a great collection of village centres all joined together. That walkable distance is the one that dictated the spiral-like basis of Poundbury's layout. The determination not to put the car at the centre of things also influenced the design of the street plan, to the extent that in many places there is not so much a road as a space. It is the buildings that define these spaces rather than a road layout that dictates where the buildings appear. So there are moments in the design process when what seems like a road suddenly opens out into a square, sometimes wide, sometimes narrow. This is exactly the opposite of what happens in every other conventional, Modernist development. Also, a rule closely observed is to make sure that there is always some sort of structural 'event' at regular intervals. It could be a fountain, a tree or a bench, but the combination of this varying shape to the roadways and surprise events results in drivers naturally going slowly. There are no traffic signs nor any white lines painted on the ground – such natural traffic-calming measures built into the design help to prevent accidents involving a pedestrian and a vehicle.

Most house fronts connect directly with the thoroughfares and public spaces which are laid out so as to maximize the flow of pedestrians. High-quality materials, locally familiar designs and the use of craftspeople rather than prefabricated units confers on Poundbury qualities of authenticity and local character. The localist philosophy has also helped to retain and expand the local skills base in traditional building methods.

An important principle that the masterplanners and designers of the houses were asked to reflect is that ancient injunction to 'do unto others as you would have them do unto you'. As I mentioned in the previous chapter, it can be very revealing and instructive to ask planners, developers and architects where they themselves live!

As climate change has more firmly arrived on the public agenda, so the more recent construction phases at Poundbury have included the development of more efficient 'eco-houses'. These do not, however, surrender to Modernist design principles, which tend to produce 'green' houses that look like spaceships. Their designs maintain the commitment to high-quality materials and traditional styles used throughout the rest of the development. The process is now being taken an exciting stage further by looking at a better way of powering Poundbury. I have been immensely impressed with the work done in Germany and Austria, where there are now around 3,500 anaerobic digestion plants. This technology provides the opportunity to convert organic material, including food waste and animal manure, into energy through the production of biogas. This is combusted to produce electricity and, in the most efficient versions of the technology, also produces heat as well as valuable fertilizer in the form of a composted material at the end of the process. It is one very effective alternative to fossil-fuel-based energy generation and such a plant is now planned for Poundbury. This will rely on a variety of waste materials to create power and, to begin with, locally grown maize as well.

Inevitably, I suppose, Poundbury has attracted the attention of critics. It seems, however, that many have deliberately chosen to ignore how the new settlement is underpinned by a radically different philosophy which, while in part inspired by traditional wisdom and experience, very much looks forward, anticipating and dealing with real twenty-first century challenges.

In many ways, Poundbury has emerged as a counter-model to the prevailing Modernist paradigm; a practical, genuine alternative to what has become a monocultural approach. It is designed for people and not cars and has been created to enhance the social and environmental atmosphere. The project tries in all it does to think about the future, which is why, for instance, what at the moment is a cricket pitch and a playing field may well one day be a necessary communal resource as land for growing vegetables.

And despite the predictions of some of the doubters, Poundbury has proved to be commercially very successful. Simon Conibeare, who has directed the development from the outset, openly admits that when I first suggested this idea, he argued that it would be financially disastrous and advocated simply selling off the land to developers to let them get on with what they have always done. Now, he is very pleased to admit that he was wrong! Property has not only held its value in Poundbury, it has increased it. Clearly, from a purely commercial point of view, this sort of development, based on the principles I have explained, is far from an indulgence of the imagination. It has proved to be what I always knew it would be: a very successful business venture, as well as enhancing social and environmental value.

Seeing this pioneering new urban extension working at most levels prompts me to ask some searching questions about why we continue to tolerate suburban developments that fail to reflect modern environmental and social concerns. They are built seemingly with no regard to the fact that they will face big economic challenges in the future, some that will threaten their viability, not least the likely volatility of energy prices.

The Natural House at the Building Research Establishment in Watford, England. This experimental building project seeks to marry traditional design and materials with the modern challenges of energy efficiency and sustainable mass construction. This work shows how we have good reason to be optimistic about how we can meet our needs for comfortable dwellings at the same time as protecting Nature.

Planning the layout of a settlement that will be able to face those challenges is, of course, not the only way that progress is being made in creating greener places to live. Building design is massively important as well. Around the world different organizations and initiatives are seeking ways of converting existing buildings and constructing new ones in ways that will be comfortable, but will also minimize emissions and the use of resources.

As I have learned more about the urgency of the ecological crisis that faces us, I have encouraged my Foundation for the Built Environment to learn more about truly low-impact design. One result is the building of a trial 'Natural House' at the Building Research Establishment in Watford, Hertfordshire. This project is looking for a new model for green or low-carbon building that can be easily adapted for volume house-building. Its design has a contemporary yet timeless feel, even though it is based on time-honoured geometric principles of balance and harmony.

Instead of bricks it uses a new design of interlocking clay blocks, which are low-fired and therefore low-carbon. They are much quicker to lay than normal bricks and are moulded in such a way that they breathe while also providing excellent insulation. The house uses lime plaster, a low-carbon alternative to cement, and sheep's wool for additional heat retention. The house will also have a heavily insulated roof made from wood fibres and hemp.

The design will soon undergo tests to assess its 'liveability' and it is but one example of the many opportunities we have to build in a more sustainable and low-carbon way, using not only modern technology, but also traditional methods in an integrated process.

To me it seems that there is much we can do to improve the well-being of both people and our planet by going about things differently in our built environments. Again, working with the principles of harmony in mind and seeing Nature as our tutor rather than a cold and boundless supplier of resources, puts us into a much more productive frame of mind. Why on Earth this should be considered 'old-fashioned' baffles me, when surely it recognizes the fundamental, timeless realities of our existence on Earth.

The design revolution

In recent years I have been hugely encouraged to see how reconnecting with Nature through alternative methods of food production and farming, by reflecting principles of harmony in how we promote health, and by the way we design buildings so as to enable communities to flourish, are all becoming more familiar in popular culture and debate. I have also been encouraged by the huge expansion of more benign technologies that could help us end our dependence on fossil fuels. Engineering and design in the clean energy and transport sectors have, in recent years, shown us the vast potential that exists for working with the grain of Nature in meeting our energy needs. From wind energy to concentrating solar power and from photovoltaic electricity to wave power, there is a vast range of design solutions out there that could help us cut greenhouse gas emissions very quickly. After a great deal of battling, even the cars I drive are now able to run on renewable energy - in the case of my 40-year-old car, believe it or not, on a waste product of wine-making.

I will not dwell on these technologies here, however, as much has recently been written on this subject, not least in *Our Choice*, by Al Gore. I met this tireless campaigner many years ago when I interviewed him for my film *Earth in Balance,* and his recent book provides a very thorough summary of the solutions we have for cutting emissions from transport and energy. What I want to do now is explain something slightly different, and to explore how Nature herself could be a powerful inspiration in facing many of the engineering and resource challenges of the coming decades. Part of the story is a hugely exciting branch of engineering based on the harnessing of solutions to problems developed through hundreds of millions of years of evolution. It is called biomimicry, which means the 'imitation of life'.

Life on Earth has been in the business of solving the complex challenges of survival for more than 3.5 billion years. It seems to me that the more we understand the innovations that have resulted from this process, the more we realize that many of the solutions that must be mobilized to meet the needs of our expanding population without destroying the natural world that sustains us, have already been invented – not by scientists in laboratories, but through the aeons of trials and tests that have taken place in Nature.

Whereas great claims have been made for 'biotechnology', and all of the potential benefits to be gained from moving genes from one species to another, relatively little attention is still devoted to how we might derive huge benefits from looking at the way Nature solves her problems and meets her energy needs. This is a fascinating subject and I want to offer just a few of the many exciting innovations that are coming out of this new blend of physics, biology and engineering.

I was told recently how Jay Harman, an Australian naturalist, found inspiration in an unlikely place: at the bottom of a sink. He had been a keen swimmer and as a youngster had dived off the coast of Australia and loved studying the coral reefs, where he saw the same thing happening over and over again. It was something he also saw every time water disappeared down the plughole at home – spirals. Harman eventually froze one of these water spirals using liquid nitrogen, then he made a mould of it and discovered that, not only does it contain those same geometric proportions that ancient scholars were aware of which frame the world's sacred architecture, but when he inserted this spiral mould into a pump, it proved to be a hugely efficient propeller. This discovery led to the development of tiny pumps that are now used in a variety of applications. They are used in vast tanks of drinking water in the US, for

Nature has been in the business of solving complex problems for literally billions of years. This beetle is using special structures on its back to harvest water from fog in the Namib Desert, Namibia, Africa. The beetle has inspired new technologies that might be deployed in water stressed and arid areas.

instance, to keep the water moving and preventing it from becoming stagnant, all using just a small amount of electricity.

Designs inspired by Nature can not only move water, they can capture it even when you cannot see it. In regions where there is little rainfall, animals and plants have evolved the means to harvest moisture from clouds and fog, even when the fog is so thin it is invisible to the human eye. One such place is the Namib Desert, where there is a creature that does this. The Namibian fog beetle miraculously manages to survive in a region of Southern Africa that is among the driest on Earth. It does so courtesy of tiny bumpy structures on its body that accumulate moisture from the thin fogs that flow over the land from the adjacent cool ocean. The beetle goes to the tops of dunes when misty breezes occur, turns its body to face the wind, straightens out its back legs and lowers its head. The mist settles on the beetle's back, where it forms droplets of water, and these, in turn, slide down so that the insect can drink.

Designs inspired by Nature can not only move water, they can capture it even when you cannot see it. In regions where there is little rainfall, animals and plants have evolved the means to harvest moisture from clouds and fog, even when the fog is so thin it is invisible to the human eye.

Biomimetic engineers have studied these beetles' bumps and come up with a much more efficient alternative to the relatively inefficient vertical netting that people already used in some dry areas to harvest water from mist. By embedding tiny glass spheres in wax to create a pattern of water-attracting peaks and water-repelling troughs similar to the ones on the beetle's carapace, they found that the surface formed water drops when exposed to a fine mist. The design can easily be reproduced on a larger scale by printing hydrophobic inks onto hydrophilic plastic sheeting and this device can capture useful quantities of water in otherwise arid areas, thus reducing the need for engineering solutions that cause damage to wetland ecosystems.

'Magnetic' termite mounds derive their name from being aligned north/south thereby minimising the surface exposed to the midday sunshine and preventing overheating. We can learn a great deal from these and other animals in the design of more efficient buildings.

When it comes to energy there is also vast potential for gain through looking at how Nature has approached a variety of problems for tens of millions of years. The landscapes of many tropical regions are dotted with myriad termite mounds. For new visitors to the tropics they are initially eye-catching structures and scientists have taken a very close look at them – and then made some amazing discoveries, some of which have already found practical applications.

One example is the Eastgate Centre in Harare in Zimbabwe. Constructed during the early 1980s, this commercial building embodies design features developed over millions of years by termites. The remarkable thing about this ordinary-looking structure is that, despite the tropical latitude at which it is built, it has no electricity-powered air-conditioning. It stays cool thanks to a ventilation system invented by the termite *Macrotermes michaelseni*. These little insects build mounds that are self-cooling and these maintain the temperature inside the nest to within one degree of 31°C day and night. This is in a country where the temperature outside varies between 3°C at night and 42°C during the day. Whereas so many tropical countries continue to mimic the steel and glass constructions typical of industrialized countries' cities, Eastgate, by using designs inspired by the termites, consumes only 10 per cent of the energy needed by a conventional building of similar size.

More recently, researchers have undertaken the digital scanning of termite mounds to map their three-dimensional architecture. Computer models will help scientists understand exactly how the tunnels and air conduits manage to exchange gases, maintain temperature and regulate humidity. This work may well provide blueprints for self-regulating human buildings in the future. Imagine what we could do to reduce carbon emissions if we combined the genius of traditional design with the genius of the termites.

Not only can Nature provide the means to save energy, it can also provide insights into better ways of generating power. I was recently very interested to hear how the insects trapped in amber that were so vital to the plot of *Jurassic Park* have also provided inspiration in the real world. Researchers at Oxford University and London's Natural History Museum investigated the eye structure of a tiny fly that had been trapped in amber for 45 million years. They discovered that the eye had special properties and reflected almost no light over most angles – apparently an adaptation to enable the fly to see better in dark conditions. By mimicking the structure of this special eye surface, engineers have developed a coating for solar panels that reduces the amount of light they reflect, thereby boosting the capacity of these 'biomimetic' panels by 10 per cent, compared with a panel without the fly-eye design.

In materials science, too, there have been some amazing recent discoveries with potentially dramatic implications. One of the strongest and toughest fabrics known to man is Kevlar. It has a variety of uses. For example, it is used to make flak jackets and extra-tough tyres. It is manufactured by pouring molecules derived from petrol into a pressurized vat of sulphuric acid, which is boiled at several hundred degrees Celsius before the fibres are forced into alignment by high pressure. The whole process requires a huge amount of energy and produces toxic by-products.

There is good reason to believe, however, that it is possible to manufacture materials that are even stronger than Kevlar, but without all the pollution. Spider silk is much tougher than Kevlar. Ounce for ounce it is five times stronger than steel. Arachnids, of course, need no sulphuric acid, nor intense pressure; spiders make silk in water at normal air temperatures. Researchers are currently studying this particular self-assembly process and their findings could yield huge dividends in terms of reducing toxic pollution and emissions.

Another creature to have excited the bio-engineers is the humble mussel. Perhaps more familiar as an appetizing meal, these bivalve molluscs exude an epoxy substance to rival the strongest human-manufactured glues. Mussels produce the sticky material in their fleshy foot and then use this to attach themselves to rocks or other submerged structures. Created in seawater, it remains intact in that environment, it is produced at low temperature (thus with relatively little energy) and is non-toxic. Although work on this substance is still at a research and development stage, engineers are confident that it could

be of tremendous use in ship-building and the construction industry, as well as in medicine and dentistry.

One more area where biomimicry could help reduce environmental pollution is in the development of alternatives to traditional paints. The next time you see a brightly coloured butterfly and give the graceful creature hardly a second thought, just remember that on its delicate wings it might carry, not just the body of an apparently insignificant insect, but also the means to do away with chemical pigments. Some butterflies contain an alternative to the kinds of pigments found in paints. Many of their bright colours are not colours at all but the result of an illusion created by tiny layers of membrane, in the nanostructures of their wing scales, that interfere with the light.

Biomimetic engineers have mimicked this structural feature to give surfaces colour without the need for chemical or oil-based pigments. The idea is to create surface layers that split white light and selectively reflect colours, rather than to use a coat of paint. This 'thin-film interference' has the potential to create colours that are brighter than pigments, that will never need repainting and avoid the toxic effects sometimes associated with the mining and refining of pigments.

Among the first products from this research are pigment-free fibres inspired by blue butterflies from the genus *Morpho*. Colour differences are created by varying the thickness and structure of the fibre that mimics the butterfly's nanostructures. Energy consumption and industrial waste are reduced because no dyeing process is needed.

Reducing the need for cleaning materials, such as detergents, is another possible way for us to diminish our demands on the Earth by harnessing solutions already invented by Nature. Lotus plants live in muddy ponds, but

ABOVE LEFT: *Some species of fly have evolved structures in their eyes that have helped to improve the efficiency of solar photovoltaic technologies.*

ABOVE: *Solar power is being deployed on an ever larger scale, assisted in some cases with inspiration from Nature.*

somehow manage to keep their leaves spotlessly clean and shiny. Engineers studying these plants have found that their secret lies in microscopic structures on the leaf surface that stop water droplets from getting a grip.

Raindrops are forced to remain as spherical blobs that roll, rather than slide, across the leaf, rather like blobs of mercury. As they roll, so they collect the dust particles that lie in their path and this and other residues roll off with the raindrops, so the leaves look clean even after being splashed with mud when, say, an animal passes by after rain. Perhaps it is unsurprising that this adaptation has led the Lotus plant to become a symbol of purity and cleanliness in some Eastern religions.

An application to have emerged from research into how the Lotus plant keeps itself clean is a kind of paint called Lotusan. It works through replicating the surface of the leaf. It is applied as a surface to Man-made structures so that when it rains the surface, for example the inaccessible exterior of a tall building, cleans itself. Algae and fungal spores are either washed off or are unable to survive on a dry and dirt-free exterior. This Nature-inspired application has the added benefit of saving the considerable costs that can come with cleaning inaccessible parts of buildings.

Many species of butterfly create colour without pigment, instead using light splitting structures held in tiny scales on their wings. The Blue Morpho butterfly has inspired technologies which remove the need for the relatively polluting processes required to manufacture chemical pigments.

One of the most difficult conundrums we face in cutting climate-change emissions, while at the same time enjoying all the efficiency and convenience of long-distance travel, is what to do about aeroplanes. At present they cause only a few per cent of global emissions, but flying is expected to expand rapidly in the next few decades so if there is to be a good chance of achieving those mid-century targets of reducing the total amount of greenhouse gas emissions by 80 per cent or more, then breakthroughs in aircraft technology will be vital. It is perhaps paradoxical that the inspiration for this comes not from creatures of the air, but from a creature of the deep.

Unlike commercial aircraft wings with a straight leading edge, the leading edge of the humpback whale's flippers is scalloped with prominent knobs called tubercles. Surprisingly, when subjected to tests in wind-tunnel experiments, the scalloped flipper proved a more efficient wing design than the smooth edges used on commercial planes. In tests comparing the relative effectiveness of a scalloped and a sleek flipper, the scalloped designs were found to have 32 per cent lower drag, 8 per cent better lift properties, and could withstand stall at a 40 per cent steeper wind angle. This remarkable discovery has the potential to improve airplane wings, the tips of helicopter rotors, propellers and ship rudders, rendering them more manoeuvrable with reduced drag, so improving fuel efficiency. The improved stall angle could improve safety margins and so make flying safer too.

Research findings could also help to boost the efficiency of wind turbines and make them quieter. A new company called Whale Power uses tubercle technology and promotes its products and ideas. I rather like their slogan: 'Building the energy future on a million years of field tests.'

In the automotive industry, too, biomimicry is making a difference. For example, Chrysler has come up with a car design inspired by the boxfish. This creature optimizes the strength of its skeleton by making the structure strongest where loads and stresses are the greatest. Vehicle designers at Chrysler developed a computer programme using the same principles to create a super-strong, but super-light car that cuts the cost of materials, improves safety and reduces fuel consumption.

Another focus of research with a potentially huge range of applications is the way in which geckos walk on vertical walls and even upside-down on ceilings. This remarkable feat is achieved with the help of millions of nanoscale hairs on the underside of the lizards' feet. These key into even quite smooth surfaces, enabling the creatures to obtain a grip that can easily support their body weight.

The surface of the Lotus leaf has properties which cause water to form droplets which roll over the surface in a manner that collects dirt and dust particles. This automatic cleaning system is now being mimicked in a self-cleaning paint. This helps reduce the need for detergents and other cleaning agents that add to environmental pollution.

Adhesive tapes that use this principle are already being produced and can create far stronger bonds than conventional alternatives.

Another idea was not only inspired by Nature, but by what was arguably her most innovative period. Some 542 million years ago there was an explosion of life on Earth, with many of today's familiar animal structures, such as shells, exoskeletons and jointed limbs, appearing for the first time over a very short period in evolutionary terms. Debate has raged for decades as to the cause of this sudden explosion of life, but evidence assembled by Andrew Parker, a biologist at the Natural History Museum and Oxford University, suggests the evolution of fully-fledged vision played a very significant part. The animal was the trilobite which had a brain and body to exploit this innovation as a fleet-footed predator. Once some animals evolved eyes, they could see other animals and then could catch and eat them more efficiently. Under pressure, the hunted animals evolved armoured parts – spines and shells – and eyes too. But not only did they need light-gathering organs, they also needed bigger brains to process the visual images that the eyes collected, thus prompting a further spurt of development.

This period of evolutionary innovation is now inspiring new software tools to help companies understand the best ways for them to optimize positive change towards more sustainable business. Data is fed into the programme and then complex scenarios are constructed around the disruptive influence of different possible changes in company practice. The emergence of the eye caused massive disruption to the evolution of life and so led to all kinds of innovations that could not be anticipated. And so it could happen with business. By modelling the effects of potential changes it will soon be possible to optimize decisions that have the biggest impact, but keep financial and environmental costs low.

After many years of investigation I am convinced that inspiration from Nature, and the way we use what we have learned in designs and technology that enable a more sustainable approach, could be the basis for a new Industrial Revolution. Unlike the coal-powered one that began in the eighteenth century, this would be part of the Sustainability Revolution rooted in building harmonious relationships with our planet's life-support systems, her rhythms and her cycles. I hope it is clear that Nature holds so many real solutions to our crisis of sustainability, if only we take the time to look for them. It seems that recent advances in various branches of science rather vindicate my instinctive feeling, but what stands in the way of their really playing a helpful role is the way business looks at them, through conventional economic glasses.

Time and time again I have found to my frustration how sustainable

PREVIOUS PAGE:
A humpback whale breaching, Hawaii USA. It is somewhat paradoxical that animals of the deep ocean should inspire technological solutions used in the air. Copying the shape of the fins on these animals has led to breakthroughs in wing design and the efficiency of wind turbines.

HARMONY

approaches, whether they be in farming, architecture, engineering or planning are so often dismissed because in the short term they are deemed uneconomic. The costs of research and development or simply of installation outweigh the benefits, it is argued, and therefore we are forced to persist with a process of industrialization similar to that to which we have become accustomed and dependent upon. The revolutions in farming, the built environment and industry that I have highlighted here and that are needed if we are to accommodate the future demands we will place on our small planet, are dismissed because they are deemed unworkable from an economic point of view – or, I should say, from the current economic point of view. That, too, needs to be reviewed.

Changing our approach to economics is a huge task, but it is a vital one and a challenge that I have no doubt we must rise to. But changing our approach to economics goes hand in hand with how we shape our cultural outlook and that means the way we educate children and adults. It is these subjects I would like to touch on briefly next, because a big part of the solution to our many entrenched problems lies not only in the kind of practical action we have just explored, but in how we view economics and education.

The boxfish has evolved the ability to create a very strong body structure using a minimal amount of material. Car manufacturers have been inspired by the fish to develop a new vehicle concept that saves metal, and thus materials and energy. A boxfish swims in an aquarium situated near a Mercedes-Benz bionic concept vehicle during an innovation symposium at the Washington Convention Center in 2005.

Foundations   6

First they laugh at you, then they ignore you, then they fight you, then you win.

MAHATMA GANDHI

In early February 2009 London was covered by the biggest snowfall it had seen for eighteen years. Transport systems ground to a halt, schools and offices were closed. Within hours the news bulletins spoke of costs to the economy of some £1.2 billion. But that was only one way of looking at things and one way to measure the snow's impact. Another, less tangible result was to be seen in the smiles and laughter of families who went out sledging, or people coming together in the parks to build snowmen; children with their parents enjoying the exceptional quiet of the empty streets and the stunning white landscapes. That the Media chose one verdict on the snow and not the other speaks volumes about how modern societies judge themselves. On that snowy day the words of Robert Kennedy came to mind: 'The Gross National Product measures everything except that which makes life worthwhile.'

It would be wrong to draw from this that I think GDP growth is a wholly redundant measurement. We have seen in recent decades enormous increases in material comfort right around the world, but as I have already made clear, there are serious questions about what is being counted as 'growth' and 'progress', especially in those societies where material needs are now for the most part not only met, but in some cases far exceeded, and where there is good reason to believe that narrowly defined GDP growth is no longer an efficient way of improving well-being or happiness.

Perhaps, therefore, there is a need to move towards the kind of economic thinking that promotes quality of life, rather than simply the quantity of consumption? GDP growth was an idea of its time, a mid-twentieth-century concept that fitted with the circumstances of the era in which it was conceived, but now the challenges are different, and maybe we need new economic tools to deal with them.

I have always been rather intrigued by how the country of Bhutan measures things. High in the Himalayas, people in this remote mountain kingdom measure well-being and progress by GNH – that is, Gross National Happiness. They happen to be largely cut off from the blanket consumerism that now dominates the Westernized world, where growth in GDP is an obsession of our

PREVIOUS SPREAD: *July 2007. Torrential rain leads to flooding that threatens to cut off the town of Tewkesbury, England. That month of flash flooding caused major disruption across England. According to insurance industry figures there has been a global trend towards more frequent extreme weather conditions.*

LEFT: *In February 2009 London experienced snow falls not seen for about 18 years. Commentators were quick to point out the financial costs of the unusual weather, as transport ground to a standstill and people couldn't get to work. What was less spoken about was the huge boost in happiness that many people felt, as they joined together in rare winter fun. Is there something fundamentally flawed in the way we judge success in most societies?*

entire culture, so perhaps Bhutan is ideally placed to conduct such a unique experiment. Certainly if you go there, as I have done, it is a magical atmosphere, so maybe there is something in it.

There is certainly increasing interest in moving the debate about how we view economic progress into a new phase – a phase that reflects the challenges of the present century, rather than those of the last one. Even world leaders are asking these questions. President Sarkozy of France set up a new commission in 2009 to look at measurements of progress.

Our propensity to concentrate on growth in this way is the same whether we are judging the national and international economic picture or when our attention focusses on the performance of our corporations and businesses. In an effort to tackle the latter element of the picture and to push forward this whole issue, I launched a project in 2006 called Accounting for Sustainability. Its purpose is to demonstrate how new methods of measurement in business and commerce can help create a clearer and more comprehensive assessment of an organization's performance beyond its simple financial results. It is all about measuring the impact we make on the environment and Nature, which is not measured in conventional accounting, and I am delighted to see how the project has acted as a catalyst. It has brought people together to develop the kinds of practical tools that will enable environmental and social performance to be much more connected with business strategy and financial performance and thereby embedded into the way a company operates on a day-to-day basis as it makes its decisions.

Among the project's many achievements so far is its work with the UK's Treasury, which now has a target for adopting new methods for reporting its wider performance. The intention is for its financial information to be incorporated within an integrated assessment of sustainability. In other words, it will list the impact its activities have on the environment – all in one report. My project team is also working with a wide range of private companies to help them develop and use similar accounting and measurement tools that I feel are absolutely vital if the Sustainability Revolution is to be at all successful. One of those accounting and measuring tools was developed with some of our leading stakeholders and is now helping them arrive at a more comprehensive understanding of their overall performance. It is called the Connected Reporting Framework. This may seem a rather dry thing to mention but, believe you me, this has turned out to be a very effective tool and it could play

Trucks line up next to a berthed container ship at the port of Keelung in Taiwan. We have become used to ever more global economic integration. This strategy for global growth and development has achieved some economic goals but has often undermined social and environmental ones.

an important role in bringing about a more harmonious practice of business – one that we clearly need to embrace.

By adopting a much more joined-up approach to the way a company reports its performance, we found that a lot of the data that organizations need is already available and, even better, the time needed to collect and present it is not too onerous. As we all know, accountancy is all about the bottom line, but there is good news here too. So far we have found that improving environmental performance saves money. It cuts the cost of energy and things like water bills. What is more, this 'whole-istic' way of reporting enables companies to carry out better strategic planning, simply because they have more data to go by and data that maps the wider implications and previously hidden costs. And there has been another very important social spin-off too. We have found that by producing these more comprehensive assessments, a

company's staff, its board members, shareholders and, indeed, its customers have all become more aware of the challenges that come with trying to act sustainably.

In more recent years companies which have signed up to the Accounting for Sustainability Project – some of the biggest in the world – tell me that by engaging with the challenges highlighted by this broader approach to assessing their performance they are enjoying real advantages in their businesses. For example, HSBC and the insurance company, Aviva, have found that their community of investors are increasingly interested in information that relates to sustainability and will more readily invest if they can see that these matters have been taken into consideration. Aviva has also found that improvements to its range of information are not only a basis for a much more effective engagement by its stakeholders, but have enabled the firm to include sustainability targets in executive bonus schemes.

It is my hope that this kind of work can help to dismiss the idea that efforts to achieve a more comprehensive, harmonious way of relating to the world around us, either personally or corporately, are some sort of vague and idealistic process. By widening the bottom line and including the numbers that refer to sustainability, it is becoming possible to show practical ways of how we might all better operate in the future.

A case in point for the financial sector, one that has been high on my list of priorities in recent years, relates to the continuing disappearance of the world's tropical rainforests. While banks and financial institutions might regard their businesses as having little impact on the environment – with their attention focussed perhaps merely on the paper and energy they use in their offices – it is the case that they are financing much of the destruction of these vital rainforests through the decisions they take about loans and investments, including those that see them investing in monocultural plantations of oil palm. This represents an economic failure of grand proportions.

Profits may accrue and loans may be repaid with interest as palm oil plantations mature, but what is not being included in the financial calculations is the loss of rainfall, the depletion of biodiversity, the release of carbon dioxide, or the impact on local communities – all of which should count as a loss or cost to society and a reduction of Nature's capital. At the moment, few companies in the financial sector are in any position at all to understand these wider impacts of their business decisions, let alone to reflect them in how they judge business success. But there are signs that this is now starting to change.

The ClimateWise initiative that I helped set up in 2007 was put together to unite a group of leading insurance companies in seeking solutions to climate

change. In conversations with these companies I had suggested to them that their sector is particularly vulnerable to the impact of climate change. Why? Because extreme events are already increasing the cost of covering losses, and this is leading to more and more properties and other risks becoming uninsurable. ClimateWise set out to promote action through better risk analysis using the most recent climate change projections, to give a clear voice to policy-makers on the need for urgent action and to support awareness among customers. There has also been work to reflect low-carbon development in investment strategies and to become better at more comprehensive reporting. This is one more practical example of how economics can connect with what is happening in the world, and to begin putting in place a different kind of relationship between all of us and the planet that we rely on for all our needs.

Natural values

As I mentioned earlier, one area where I have believed for a long time that a new approach to economics is needed to solve a major ecological challenge is the plight of the rainforests. This is why I established my Rainforests Project in

Bush fires rage across South-eastern Australia. Some climatologists say that a long-term regional climate change has taken place, with drier conditions now the norm. The corresponding increase in the risk of fire has prompted the insurance industry to respond with increased premiums or withdrawal of cover, holding major implications for businesses and people.

late 2007 with the aim of trying to find a solution to the rampant clearance of the tropical rainforests, based on a very straightforward, simple idea, that the trees are worth much more alive than they are dead. As I explained in Chapter 2, they provide services of untold value to the rest of the world. Not only do they pump billions of tonnes of water into the atmosphere every day, thus maintaining rainfall over a wide area, they also sequester vast amounts of carbon. They do this naturally and free of charge. The alternative is to use very expensive technologies such as carbon capture and storage (CCS).

It had become very clear to me that deforestation is predominantly an economic problem and one that requires predominantly economic solutions for it to be fixed. I saw that through my Rainforests Project I was able to bring together a number of heads of government from around the world to discuss the issue, and that led to a group of countries, with Norway acting as the secretariat, meeting to identify how new financial mechanisms could be deployed. The aim being to turn the forests into an economic asset – the absolute reverse of the situation that has prevailed throughout history, whereby countries pursue the obvious incentives to liquidate their forests for financial reward. Proposals were developed, essentially to compensate countries for departing from their historic deforestation patterns, or for not beginning large-scale deforestation in the first place.

The sum needed to help the rainforest nations do this was estimated to be about 20 billion Euros over the period 2010–15. This would be sufficient, it was estimated, to cut deforestation by around a quarter, thereby saving emissions of about seven billion tonnes of carbon dioxide – or about the same as the annual emissions of the USA.

While 20 billion Euros is certainly a significant sum, in the light of recent events in the finance and banking sectors I am inclined to see it as a rather modest amount, especially as it will help protect rainfall and thus ensure future food security. At the same time it would deliver highly cost-effective cuts in greenhouse gas emissions. To put the figure into perspective, perhaps it is worth mentioning that the finance needed to cut deforestation by this significant amount is about the same as was reportedly paid in bonuses during 2009 to the staff of Goldman Sachs.

While there is still more work to do on this subject, there has happily been some progress. For example, the governments of Norway and the rainforest country of Guyana got together in 2009 to establish a partnership whereby Guyana will be rewarded by Norway for keeping its extensive forests intact. I was delighted that my Rainforests Project was able to play a helpful role in facilitating this outcome and I hope that in the years ahead we will see a great

deal more of this kind of practical cooperation, which, I might add, now extends beyond government. We are also working with a group of pension companies to see how possible it might be to make the investment in so-called 'Green Bonds' an attractive option, and I hope that this may lead to a substantial investment in low-carbon development projects, including those that lead to a decrease in deforestation.

I have learned from my many travels how, if the resources needed to save the forests are targetted correctly in helping to empower local communities to create the conditions for change, it can have a major positive impact on poverty, on standards of health care and on sustainable farming. The money would also pay for the restoration of the vast areas of land that have already been degraded through deforestation. Estimates vary as to how much degraded land there is in rainforest regions; the commercial arm of the World Bank, the International Finance Corporation, estimated in 2009 that there are up to 96 million hectares spread throughout Indonesia and its islands. In Brazil, degraded land that could be used for productive agriculture is equivalent to the size of California, over 60 million hectares. If this land could be brought back into production rather than destroying yet more pristine forest, it could improve the economic

More public awareness about these big issues is, I think, absolutely vital to making successful change. This is why I decided to launch a campaign using a computer animated frog to encourage support for the proposals of my Prince's Rainforests Project.

Clouds of white smoke mark the extensive tracts cleared to make way for oil palm trees. Palm oil is used in products from chocolate bars and breakfast cereals to shampoo. As demand rises, more land is cleared and more species are pushed towards extinction.

development of such regions, reducing poverty as it aids the ecological recovery of the land. Taking an agro-forestry approach, for instance, could make a huge potential difference to the health of the biodiversity of such areas. Redesigning palm oil and soya plantations so that they incorporate mixed forestry would create opportunities for jobs and community participation while at the same time creating vital 'biodiversity corridors' that could reconnect the surviving, often isolated, parts of the rainforests. In fact the opportunity we have to save the forests could be a powerful example of the kind of vision for the future we might embrace; one in which we invest in both the community capital and the natural capital on which we all depend. We must go beyond seeing it as an either-or choice. Such a whole-istic world view that seeks to work with Nature and people is surely what we must now strive to teach at every level.

I am especially encouraged to see that some leading firms are beginning to respond to the challenges posed by continuing rapid tropical deforestation. The two UK supermarkets, Sainsbury's and Waitrose, together with Unilever, are among a growing group of companies in the global supply chain for palm oil which are taking steps to ensure the material they use comes from more sustainable sources. As global demand has increased, so more forest has been cleared to make way for yet more plantations of oil palms, but by setting clear criteria for more sustainable produce it is possible for retailers and food processors to make a positive difference.

It is a big step, but one that must be taken, and I am delighted that more and more companies are getting involved in this way. They have decided to accept responsibility for the impact they help cause, and to do what is necessary to minimize it. It is not simple, and there is no quick fix, but these companies have decided to take the first steps on a critical journey. As a result, they are emerging as true pioneers and leaders and I have an instinctive feeling that, in the end, this will not only be good for the rainforests, but good for business too.

For others to follow quickly, though, more incentives will be needed. Not all companies are so enlightened and they will need encouragement as well as good leadership to make the transition already being made by the minority who have understood the implications of what is happening, and that means the active support of governments. Well-targetted incentives or disincentives can play a major role. So, too, can effective partnerships between the public and private sectors and the many non-governmental organizations. Ultimately, one of the biggest influences on the way economics operate is decisions made by governments on things like tax policy, procurement, spending priorities, what kind of research they back, the trading relations they establish and the market incentives they introduce. All these tools can be used to create more sustainable economic activity.

Economics is a lot to do with the relationship between supply and demand but this dynamic has been shaped in a remarkably positive way recently by various groups who, with small resources, have managed to make a very big impact. Take ForestEthics, for example, which was born out of a protest against logging in North American forests. The campaign had failed to stop the logging by protesting in the forests, despite what turned out to be the biggest mass protest and civil disobedience in Canadian history. Only when the campaign turned its attention on the companies who bought the wood pulp and used it to publish their newspapers or for their own products did the tide turn. They held very publicly to account organizations like the *New York Times*, as well as companies like Pacific Bell and Victoria's Secret. This put the logging companies face to face with their customers who, with ForestEthics' help, persuaded the loggers to end the clear-cutting of forest. So far this little organization has managed to secure the future of more than 13 million acres of endangered forest and another 55 million acres are on the way towards gaining

permanent legal protection. ForestEthics focuses primarily on North American forests and has taken on some of the largest companies on Earth – among them Dell, Office Depot, Williams-Sonoma and Whole Foods. It challenged Staples, the world's largest office supply store, for instance, to move towards using certified and recycled paper and away from material derived from endangered forests. When it did this, ForestEthics had an annual campaign budget that was less than the amount made by a single Staples store in a week. There were over 2,000 Staples stores across the USA, but ForestEthics prevailed and the company changed its policy.

I mention this very effective non-governmental organization because it is a perfect example among thousands of others of where small groups of people have become vital actors in the economics of positive change. These groups have to build their momentum and increase the support they enjoy so that their impact can be greater. The world really needs them, even though all too often it seems that many of us are not prepared to hear their vital message.

Crisis or opportunity?

Some argue that, at a time of financial crisis, when share values are tumbling, companies are closing down and markets are collapsing with the result that jobs are being lost and homes repossessed, this is not the correct moment to re-evaluate our economic priorities. But then again it might be exactly the right moment to raise such questions, complex and challenging though they sometimes seem. I wonder if we might see wisdom in the Chinese word for 'crisis', which also means 'opportunity'? It is almost a knee-jerk reaction to try to put things back as they were when we hit such a crisis, but we could be more positive and forward-looking and consider another approach that could create something different – a framework that is more sustainable; an approach to economics that is more proportionate, balanced and connected; an economics that recognizes the need for harmony. So what might that framework be like?

As I hope is now clear, it is not simply a credit crunch that we have to address, for other crunches are all making themselves known simultaneously. There are climate crunches and crunches to do with the use we make of the Earth's natural resources – those many 'ecosystem services' provided by the Earth's living life-support systems like forests, wetlands, river catchment systems, peat bogs and tundra, as well as marine ecosystems like coral reefs, estuaries and vulnerable parts of the ocean floor that have been degraded by deeply destructive bottom-trawling. But perhaps one of the biggest crunches of all is the fast-increasing human population. None of these problems can be dealt with in isolation, and putting the financial and economic system back on its feet without seeking a positive contribution towards solving these related, parallel challenges would be to stick our heads in the sand. If we do ignore these related issues then two major problems in the world could become even worse. Firstly, Nature's capital resources will be eroded even more than they already have been and, secondly, those who pay for those losses will fall even further into debt. It is a link not made as obvious as it should be, that when nobody pays for the loss of Nature's resilience, it is the poorest people in the world who always end up carrying the debt. They are the ones who have to struggle when there is less and less fresh water available, when food becomes scarce because harvests fail or when their homes and lives are devastated by serious storms or floods. Making the losses to Nature more visible on the world's accounts would be a move in the right direction in alleviating some of the worst of the poverty that currently besets many hundreds of millions of people.

Having talked this situation through with economists and business leaders on various occasions, I now see very clearly that, although it will undoubtedly

be tricky to get right, it is, at the moment, just possible that we could get it right. There is a small window of opportunity, but it is beginning to close rapidly. That is why now is the right moment for a more joined-up response. Perhaps it could be inspired by the principles of harmony, because the solution is all to do with finding a path that puts Nature and her virtuous circles back at the heart of things rather than on the periphery. Let me explain.

The root problem with our present economic model is pretty straightforward to understand. As it stands, it maintains a country's economic stability by increasing the production and consumption of goods and services. Economists call this 'consumption growth' and only when it is rising do politicians and economists consider an economy to be healthy. They worry when this growth starts to decline. When we have a recession, for instance, their priority is to kick-start growth so that we see a return to spending in stores. As I have already pointed out, this approach excludes the impact such an emphasis has on natural resources and the increases in emissions that compound our present environmental problems. If you take a look at what the fathers of this system had to say about continuous growth you may be surprised. People like John

Stuart Mill in the nineteenth century or John Maynard Keynes in the twentieth both foresaw that a time would come when endless growth would no longer be either necessary or prudent. Mill talked of an economy eventually moving to a 'stationary state of capital and wealth' and Keynes likewise expected a moment to come when we could 'prefer to devote our further energies to non-economic purposes'. In both cases they acknowledged that this could only happen once a certain standard of universal health and welfare has been achieved. This could be said to have been achieved in parts of the Western world though not, as yet, elsewhere.

There is the danger, of course, that if you decide to try to achieve this state too quickly and move to an approach where endless growth is no longer the priority, all sorts of collapses and even disasters can occur. Businesses go bust, unemployment rockets and people fall into the poverty trap. It may seem to those of a more radical disposition that we should just go all out to establish this new state of things, but that will only lead to the total meltdown of society. Clearly, guiding the engines that drive social and economic well-being has to be done very carefully, so perhaps the first stage might be to start thinking seriously about the sort of economic model we could possibly adopt that produces a sustainable system without all of the crippling fallout.

Tim Jackson is Professor of Sustainable Development at the Centre for Environmental Strategy at the University of Surrey. He calls this elusive model a 'macro-economics for sustainability' and, although that is a bit of a mouthful, it is a subject I would just like to unwrap a little more before we move on,

Bronze sculpture at the Franklin Delano Roosevelt Memorial, Washington DC, USA. This powerful work portrays the misery of the breadline, depicting the despair of the Great Depression.

because if we get this right we will start to move with confidence in the right direction.

As part of an excellent book called *Prosperity Without Growth*, Professor Jackson carried out a review of the research being done into this new kind of economics and was rather staggered to discover that there is little work being done on it at all. In his view that needs to change, and fast. What is quite clear from his research is that whatever model we do adopt it will always have the same aim, and that aim will jangle badly in many modern ears. For it will strive to establish an economy based upon slow growth.

One very interesting piece of work Professor Jackson did come across was produced in Canada, where researchers created a computer model of a country's economy – they chose Canada's – and then played about with those elements that drive growth. They tinkered with the figures for output and consumption, the levels of public spending and investment, the amount of trade the country might carry out and how much employment this might achieve. Then they pressed the button to see what would happen over the next thirty years. They kept adjusting the balance to see whether there is, in theory, a combination that might stabilize an economy, one that is sustainable in the long term, works within the limits of what Nature can provide and avoids the sorts of social catastrophes we all fear – high unemployment and consequent poverty. The answers this modelling came up with are complex, but they did suggest that there is potentially a route. The trouble is, it requires a very big shift away from the way things are done now, involving changes to working patterns and the way in which the various resources needed by everyone are managed.

These suggestions back up a recent report by the New Economics Foundation in the UK which concluded that continuing global economic growth is not possible if countries are to tackle climate change successfully. As it stands, the level of economic growth could not meet what it called 'unprecedented and probably impossible' carbon reductions to hold average global temperature rises below 2° Celsius. They suggested that there are no proven technological advances that will allow us to pursue that 'business as usual' outlook. In fact they said that the so-called 'magic bullets' of carbon capture and storage, nuclear power or even geo-engineering, are potentially 'dangerous distractions' from more human-scale solutions. Their policy director concluded that 'we urgently need to change our economy to live within its environmental budget'.

Professor Jackson's team looked at the question of carbon emissions and highlighted in their research another simulation model from an Italian team of economists that looked specifically at the challenges a country would face if it

made a large-scale shift from fossil fuels to renewable energy. Any such move, of course, would require substantial investment and at the right rate. Too slow and fossil fuels dwindle, causing fuel prices to rocket and economies to crash. Too fast and it could slow an economy down and dry up the resources we would need to make further investment. The researchers described this sliding scale as 'a narrow sustainability window' that can be widened, they say, if the balance between consumption in a society and investment in the economy is changed – in particular, if more national income is allocated to investment rather than to consumption. The higher this investment goes, say the researchers, the more flexible a society would be as it sought to achieve the necessary transition. But this investment would not be the kind that creates innovations that stimulate more consumption. Instead it would be investment in those new kinds of infrastructure that will allow us to move to renewable sources of energy while also safeguarding the stability and well-being of communities and society as a whole.

This is why the calculation of new measures of economic success, based more on well-being and even happiness, are rather important. If all we calculate is how many resources or how much energy is consumed, or misleading measures based on the flow of finance, we remain disconnected from the real economy of Nature and of human well-being. At the moment the importance of Nature's

Green amid the skyscrapers. Central Park, New York City. People gain huge direct benefits from contact with Nature. In urban centres, areas of parkland can add greatly to not only quality of life, but also health and longevity. But how can we measure these benefits in economic terms?

The Large Hadron Collider is a 27-kilometre-long underground ring of superconducting magnets. This incredible research project will help reveal the secrets of the Universe, but will it enable us to re-find our place in Nature?

reserves is not sufficiently visible and it needs to become much more so, in order that the connection is clear for us all to see between human health and welfare and the long-term security of Nature's capital resources.

On a recent visit I made to Berlin to receive the German Sustainability Award, I was struck by how that country has seized on sustainability as a business opportunity. In a country that has already exceeded its greenhouse gas emissions target set by the Kyoto agreement on climate change, there are now nearly two million people employed to provide environmental goods and services. Having now set themselves the target of reducing emissions by 40 per cent, the Germans expect to add another half a million jobs to that total. I think this is an encouraging example of how it does not have to mean a choice between, on the one hand, protecting our planet's life-support systems and, on the other, creating jobs and securing innovative and balanced economic development that reflects the urgent need to live off the income we derive from Nature rather than eating into her rapidly depleting capital. We also need to think a little more widely about the other types of capital that we must nurture in meeting the challenges of the twenty-first century.

What I mean by that is paying a lot more attention to the health and vibrancy that we derive from something I mentioned in the previous chapter, our 'community capital'. That is to say, the networks of people and organizations that in some cases now extend, via the internet, around the globe. But it also includes the post offices, the pubs and bars, the churches and mosques, the temples, village halls and community centres – the local wealth that holds communities together and enriches people's lives through mutual support, love, loyalty and identity. The value of this kind of community capital cannot be overstated.

In so many ways we are what we are surrounded by, in the same way as we are what we eat. So, congenial, attractive public spaces have a huge role to play. They are not just some sort of social safety net that catches those people who cannot afford their own gardens. They must be seen as opportunities for people to connect with each other and with our shared inheritance, not to mention our common future. Only if the fabric of community is strong and healthy can people engage in a common endeavour; and it is this 'fabric' that can be so enhanced by the very nature of the built environment. This is precisely why I have spent so many years working with local communities on the vital importance of whole-istic planning, design and construction. And yet, just as we have no way at the moment of accounting for the loss of the natural world, contemporary economics has no means of accounting for the loss of this community capital. But it needs to, if we hope to overcome the many practical, environmental and economic challenges that we can no longer avoid facing.

Perhaps now really is the moment to change our thinking, so that we can assess not only costs, but quality; to shift from our obsession with competitiveness and move towards durability and economic resilience; to measure sustainability as well as GDP and to elevate the emphasis on well-being rather than simply growth. Perhaps if we did this it would be possible to accelerate more rapidly our departure from the Age of Disconnection and to enter what could be the next historic phase: what we might call the Age of Harmony – certainly of Integration.

History shows that we can change, and change rapidly if we need to. Did you know, for example, that during the Second World War the UK diverted around 30 per cent of its GDP to overcoming the horrors of that conflict? Even during the relative calm of the Cold War the USA spent approximately 10 per cent of its GDP winning a decades-long arms race against the Soviet Union. Yet today, when we are far wealthier, many claim that the modest, estimated 2–3 per cent of GDP that we need to spend to slow carbon emissions to a point that might avoid catastrophe is too costly. This is yet more evidence of the deep crisis of perception I have been trying to define. Surely it is a failure of monumental proportions to see what is really important? But why do we have to wait until we are overtaken by the inevitable catastrophes that this failure will cause? If we continue to be deluded by the increasingly irresponsible clamour of sceptical voices that doubt man-made climate change, it will soon be too late to reverse the chaos we have helped to unleash.

It has to be possible to stimulate the required shift in perspective. I can say this having seen for myself the incredible level of activity in many spheres – from farming to economics – happening around the world, albeit mostly on the periphery, in the realms of the 'alternative' rather than the mainstream. But at the heart of this process of change and transformation lies the crucial issue of how we teach a more rounded view of progress.

Teaching and ideas fit for purpose

The leading anthropologist, Gregory Bateson, put it rather well when he pointed out that 'the major problems in the world are the result of the difference between the way Nature works and the way man thinks'. One person who saw this very clearly and worked to find ways of helping people think in different ways was E F Schumacher. Despite his background in mainstream economics he became one of the twentieth century's most influential figures with his powerful new ideas on how to achieve more sustainable societies. His contribution was especially surprising considering that he spent two decades working as Chief Economic Advisor to the UK's National Coal Board. His famous 1973 book *Small Is Beautiful* became a widely cited source, while his ideas for appropriate development and intermediate technology found broad resonance with the newly emerging environmental and development movements.

One of his other great legacies is Schumacher College, near Dartington in Devon, England. This is perhaps the type of educational establishment that we will need many more of in the years ahead. The college runs a wide variety of courses that unite mainstream science with a whole-istic approach to sustainable living. It pays careful attention to building a sense of community and places spiritual learning at the centre of its philosophy. The college is one of the very few educational establishments to offer an MSc in Holistic Science.

Teaching at Schumacher recognizes that the challenge we face is in many respects the crisis of perception I have outlined, as much as it is a need for different policies and new technologies. As well as building on Schumacher's legacy of thinking about a sustainable form of development – something that is imbued in every aspect of how the college functions – there is an ongoing effort to recapture the crucial balance between the rational and the intuitive in education. The staff teach the steps we need to take to nurture the intuitive and spiritual dimensions of our humanity, not to crush them. In other words, to work not just with the brain but with mind, heart and hand.

As a way of explaining what I mean by this, let me describe the work done by my own School of Traditional Arts. This is an institute in East London that I set up in 2004 and that had grown out of a previous organization I had rescued some twenty years ago that specialized in maintaining the living traditions of Islamic art, architecture and craftsmanship, as well as those of other great religions and cultures all under threat of extinction. It offers practising artists the opportunity to undertake research at the

Participation in art can create really transformative results for those taking part. Here I am visiting the College Of Traditional Islamic Arts in Amman, Jordan, and carving part of the pulpit for the Al Aqsa Mosque in Jerusalem.

highest level and study for postgraduate degrees. Students now come from around the world and they all share a desire to rediscover the values of the traditional arts and want to make a practical contribution to their survival and development within a contemporary context.

I am particularly proud of a series of projects the School has run in a town in the North of England. Burnley, in Lancashire, has suffered in the past from clashes and a breakdown of trust between sections of a community that includes various ethnic minorities but is predominantly Muslim. These clashes have sometimes been violent and have escalated into riots. Project workers from my School, operating as part of a concerted effort by several organizations and charities, have been part of the process of rebuilding the community, running week-long workshops with children drawn from the different communities and schools. The specific aim has been to create artworks and decorations for new faith centres within the new schools being built in the town. The team do this by revealing to the children – and indeed their teachers – a different way of working with the arts, one that integrates these with maths, geometry and science so that the children learn in a different way.

I have seen for myself how art can help to heal fractured communities. In this picture I am at Turf Moor School in Burnley, Lancashire, England, to meet schoolchildren taking part in anti-racism and citizenship activities. My School for Traditional Arts has been active in the town, helping to rebuild a sense of community through art.

LETS KicR Racism Out of the WOld

All of the teaching is contextualized. That is to say, instead of being taught geometry as an abstract concept that has no bearing on their everyday lives, they learn an element of the subject and immediately use it to create artworks. The entire process is couched in the context of how geometry appears in Nature and how it is used in architecture to create structure. The teachers learn as much as the children, particularly from the teaching methods the project workers use. Rather than treating each subject as a separate matter and approaching things in a mechanistic way, the project leaders always talk in terms of the symbolic meaning of the patterns they are dealing with and how everything relates back to the order found in Nature. So, for instance, when they draw a circle, they do not explain it as being a simple, geometric shape with a centre point that is always equidistant from the line and so on. They explore it in terms of a shape we see and experience all around us in Nature – in the rising and setting of the Sun each day, for instance. They ask the students what other circular forms they see in Nature, how these forms have been used symbolically by different traditions, and also what it means to be centred within themselves.

The project workers deliberately avoid telling the children that they are studying mathematics until the end of the workshop and they find that the children are delighted to discover that they have done something that previously they thought they could not do. I have seen the children's feedback to the project and many of them say much the same thing, that nobody has ever told them before that such things as mathematics and science could be so joyful and revealing. They learn what they would learn in a conventional geometry lesson, but within a symbolic context, so they relate the knowledge that they gain to their own personal experience and, crucially, they remember it.

I gather that the effect is quite infectious. Not only the children, but also their teachers who help during the workshops say that by mid-week they cannot stop counting things. They have begun to see that there are very special, universal patterns everywhere in Nature that are just as apparent in their own bodies, and it is not long before they begin to question what these mean to them and what such patterns say about their personal relationship with Nature. By the end of the week, the project workers report, children and teachers alike are beginning to place themselves within a more cosmic relationship with Nature, instead of seeing it as something separate from them and just a thing to be used. There is a genuine excitement that they are connected to everything in a very profound way.

They have begun to see that there are very special, universal patterns everywhere in Nature that are just as apparent in their own bodies, and it is not long before they begin to question what these mean to them.

I should add that working in this manner breaks down the social, cultural and peer group barriers between the different communities. The children form new friendships and work closely with each other to create the decorative tiles of a floor or the elements of a stained-glass window or whatever it may be. It stands to reason that if you are encouraged to ask each other why you might use a particular symbol or a particular colour and are talking with someone who cares as much as you do about getting it right and making something beautiful, then boundaries are bound to dissolve and the bonds of relationship are bound to become firm. My hope is that, in time, such experiences will enable these children themselves to help heal and rebuild the cohesion of what had become a divided and fractured community.

Having encouraged and witnessed these workshops from afar, I firmly believe that this idea could be applied in a much more continuous way in all schools.

This school in West Sussex, England, was inspired by an approach in Sweden where children are issued with waterproof romper suits so they can play outside even in torrential downpours. In braving the elements they learn about the great outdoors. Many of their regular lessons are conducted outside.

If it so inspires and engages the children in a week-long workshop, what, pray, is the problem with it? There is certainly no problem with this approach in one school I have learned about in Telcos at the foot of the Caucasus mountains in Russia, where children immerse themselves in studying one subject at a time until they have a very clear and rounded understanding of it. Only then do they move onto the next. In this particular establishment the children also teach each other. It is not uncommon to find a 10-year-old teaching an 18-year-old about atomic physics. The children have also helped to build the school; they do the cooking and the cleaning, and all of their lessons involve contact with Nature, so that they come out of the experience with a range of very practical skills as well as an integrated, whole and unified understanding of how the world works in all its spectacular variety.

I have great admiration for the expert work that is done in enabling students to gain detailed information about a large and growing number of increasingly technical, specialist subjects, but in the process I feel that societies are losing resilience as knowledge is emphasized over wisdom, and as specialism is promoted over whole-istic thinking. There is a danger that we end up learning more and more about less and less. This is no reflection on the professionalism and commitment of teaching staff, but more a consequence of the system and the values that underpin the criteria of success, so often measured numerically against specific targets and on the basis of how much information is transferred and remembered.

Schools are one of the main sources of the values that societies ultimately reflect, so they will clearly have a vital role to play in the kind of transformation I am talking about – perhaps even going as far as to consider stressing that the noblest aim in life is to produce the greatest happiness and the least misery; that service to community and doing work that is useful to society is as important as the individualistic pursuit of personal benefit.

It seems to me that there is a link between how a society teaches its children and that society's outlook on the world. I am sure that the values-free transfer of information that has increasingly characterized education in the West has helped to create the spiritual void that has opened the way for what many people see as an excessive personal focus – the belief that things will work out all right if everybody looks after themselves. Academic research on well-being and happiness has shown that unselfish people are typically happier. It seems quite logical to me that we should promote values of service and community in our educational system rather than simply concentrate on the skills that focus on personal advancement through the acquisition of specialist knowledge.

Recent research from the New Economics Foundation found that, once basic needs for food and shelter are secured, our happiness is fulfilled through five factors. One is our connections with other people – friends and family and the people we work with. Another is activity and exercise. Then there is appreciating beautiful surroundings and reflecting on such experiences. Another is continued learning, novelty and meeting new challenges and, finally, we gain well-being from giving and from being a member of a community. As far as I can see, at the moment we need to work all this out for ourselves; it is not necessarily what is taught as any kind of priority in the overall educational experience. If it were, perhaps the process of positive change would not be so slow and difficult.

Federation of City Farms & Community Gardens has done a wonderful job throughout the UK to support, represent and promote community-managed farms and gardens. Here I am making a visit to the Hackney City Farm in East London to celebrate the work of the federation.

While some might argue against an education and an economics system that recognize non-market attributes of human happiness, I have come across plenty of people who see the need to challenge in a fundamental way our dominant attitudes and philosophy; to ask whether they are actually fit for purpose; whether our collective world view enables us to see things as they really are. If we can find ways to ask these kinds of questions by how we deliver education, perhaps we can guide future generations towards a more integrated and wholeistic way of living; one that places greater emphasis on our place in Nature and the world.

From what I have seen, it appears that an additional part of the answer lies in restoring the contact that young people have with Nature. In 2009 the most popular place for children to play was in the home, especially with computer games. However, research by Dr Aric Sigman found that children who have contact with Nature score higher in tests of concentration and self-discipline than those who lack this and that exposure to natural environments improves children's cognitive development. Schools with outdoor education programmes were found to achieve better academic results and have better behaviour in the classroom.

One way of connecting children with Nature is school farms. Alas, many of them have closed in recent years, but one that survives is at Oathall Community

The headquarters of the World Bank and International Monetary Fund, Washington DC. These and other influential economic institutions must understand and act upon the challenges posed by sustainable development. Although making some steps in the right direction, they have a very long way to go in matching development goals with the capacities of our small planet.

College in West Sussex. I have great admiration for this farm, which some years ago was due to be shut down, but was saved at the last minute after the children wrote to me in despair. The activities of the farm are woven into every part of the college's curriculum – maths, art, biology, business and environmental studies – and so the farm is part of every child's experience at the school. For those who find academic studies more difficult, practical skills are developed and are equally celebrated by the school. Likewise, children with learning difficulties respond incredibly well by working with the farm animals. The great thing is that every child can be a success at something and this gives an enormous sense of self-confidence and self-worth. There are now twenty-five schools in England actively developing or considering a school farm. Experience of farming can be life-changing for children, as many adults will testify – especially as it connects them with the soil, where their food comes from and with the way Nature works. Not only that but, as I mentioned earlier, research by the Royal College of Physicians has shown how the Western world is now confronted by a growing epidemic of allergies, many of which can be attributed to a lack of exposure to farm animals. Given the rapid expansion of cities and conurbations it is vital that we work to give our children access to natural ecosystems and the biodiversity on which our health and well-being – particularly our psychological or spiritual well-being – depend.

All of these examples reflect the same key principle that we have to hang onto. In their different ways the institutions I have mentioned are all conscious of the risk of teaching out of us – not to say, deliberately excluding – the intuitive aspect of being by concentrating purely on the rational. The important thing is to connect pupils to the whole subject through greater rigour so that they grow up having access to an entire body of knowledge and wisdom. In other words, the best of what humanity has thought, created and written, rather than fragments of information that give no idea of the unity of things. And this should go not just for schools and other educational institutions, but also in business training.

I am pleased to say there are some signs of encouragement here too. The World Bank, for instance, is now asking many of its own leaders to undertake an intensive course run by an organization that I have worked closely with since it was founded, the University of Cambridge Programme for Sustainability Leadership (CPSL). Those taking the course study ways to build new ideas and priorities into the World Bank's mission and purpose – by no means an easy task, but certainly an essential one, and indeed showing some signs of success as it equips senior staff with tools and ideas for what might be the start of what, I pray, will ultimately be a profound transformation.

Through my different activities I have seen how more and more businesses are demonstrating inspiring examples of leadership. Nearly twenty years ago I started a series of awareness-raising seminars for business executives as part of my Business and the Environment Programme which, I am heartened to say, has so far persuaded around forty (and growing) leading pension fund companies to change their thinking and their practices and has inspired the activity of my Corporate Leaders Group on Climate Change. The journey begins by raising awareness and, by following a vision, business leaders are encouraged to transform wisdom into action. So far, thousands of executives have studied with the CPSL which remains a pioneering institute. This is undoubtedly one of the contributory factors behind the revolution in thinking that is beginning to take place among at least some companies, and that continues to gather pace.

The power of awareness and practical advice can also be seen in some industrial sectors where even quite modest changes in behaviour can deliver dramatic, positive benefits in a very short time. One, I am very pleased to say, is long-line fishing.

A pioneering scheme that unites conservationists, the fishing industry, and the South African government in what is called the Albatross Task Force has brought about a rapid large-scale decline in albatross deaths caused by long-line fishing boats. I set out earlier how these birds are in danger of extinction because of the massacre caused by long-line fishing. And yet, between 2006 and 2009, the number of birds accidentally killed by those long-line boats participating in the new scheme plummeted by 85 per cent. This incredible gain was achieved through simple changes to the methods used by boats fishing for tuna and swordfish.

Albatross Task Force instructors spend time with fishermen showing them how to prevent the birds from getting hooked. Not only do the teams of instructors train fishermen, they also investigate how to improve the design and performance of the methods used to prevent the hooking of birds. The scheme has expanded in South African waters and is now getting going in South America.

I hope I have been able to show here that as well as plenty of practical examples of how else we can go about things, there is a tremendous groundswell of thinking about how to encourage the sort of philosophical changes to the way Westernised societies are going to have to think if we are to shift from an economic system predicated on the pursuit of unlimited growth to one that seeks sustainable, durable economic development. A big part of this is of course about how we choose to live.

Wandering albatrosses feed in the vicinity of a toothfish longliner working in Falkland Island waters. While albatross populations have been decimated across the southern oceans by longliner boats, the Albatross Task Force has demonstrated through a project in South Africa how fishermen can be quickly shown new methods that enable them to continue catching fish without killing the birds.

Culture

Education is certainly a part of the process of positive change, but so too are the changing values of societies and how people choose to do things. I have been consistently impressed by different movements such as Transition Towns, Friends of the Earth and many others in how they have succeeded in spreading awareness and inspiring new ways of thinking and doing that are fit for the modern age. But there is so much more to do.

That is why in September 2010 I launched a new initiative in the UK called Start. The idea is simple: to find ways to encourage everyone to make those initial steps toward more sustainable lives. Instead of imploring people to stop doing things, to cut back or make some kind of return to the Dark Ages, I set out with the aim of making this project very much about the positive benefits that accompany more sustainable living. I sense that for many people there is a tendency to hear only a message of doom and gloom when presented with some of the issues I have talked about in earlier chapters. This is far from motivating for the vast majority of people. It seems to me that, alongside increased awareness about the reasons for change, a message of hope and optimism will win the day. I want to encourage everyone not only to see the threat, but also the opportunity.

Part of the Start plan has been to team up with some of the major companies that speak to millions of people every day and to encourage them to signal a different kind of consumption. Through the Start project I am also encouraging

I have been pleased to see the organic farming movement taking off right across the world. From India to the USA and from Scotland to Ethiopia, there are a great many inspiring examples of how we can approach food production with more sustainable methods. Here I am talking with Warren Weber when my wife and I visited his organic farm near San Francisco, California, in November 2005.

companies to cooperate with one another in bringing about the positive change that we all need. Competition takes us so far, but collaboration is also needed across sectors of industry to solve the complex challenge at hand. All this is about culture and how we do things. I believe it is possible for a more collaborative approach to be adopted between businesses. In time this could make a huge difference to our ability to make rapid progress towards aligning our needs with Nature's limited capacities.

And while it is essential for big companies to become very much more a part of the solution, it is vital not to forget the inspirational leadership being shown by small firms too. A case in point is Abbey Home Farm. In 1990 Will and Hilary Chester-Master decided to buck the prevailing trend toward ever more industrial farming methods and to turn this farm over to organic production. The couple's search for more harmonious methods was from the outset not only a more sustainable method of farming but also a deliberate attempt to inspire changes to how people think.

Above a door in one of their farm buildings is a picture of Mahatma Gandhi – Hilary says his image serves as a reminder of his advice to 'be the change you want to see in the world'. That change is not only about agricultural methods

Abbey Home Farm also demonstrates a different kind of economy. On this particular farm around thirty people are employed in tending its 1,800 diverse acres – compared to the one or two people needed to run one of the similar-sized farms nearby that produce intensive monocultures of grain. The farm also runs an apprenticeship scheme, to help train the next generation of organic farmers, the people who will be able to help in the transition to the sustainable food economy that – one way or the other – we must soon adopt.

Will and Hilary are also making intensive grass-roots efforts to educate young people in a fuller understanding of food as well as the link between land, growth and nutrition. They offer children the chance to plant seeds, nurture plants and harvest produce – and then to eat it. While industrial farms are deemed by some to be 'efficient' in generating large yields, the experience at Abbey Home Farm shows how an organic model can be much more efficient in generating social and environmental goals that societies must also reach.

Garden Organic is another organisation that encourages children to grow their own food. They have worked with support from my Duchy Originals food company to find success in even the most unlikely of inner-urban settings. They have found, as I have, that when the opportunity to see things differently is

I am always heartened by how readily children enthuse about growing food. In this picture I am on a visit to Hotwells Primary School in Bristol, England. This pioneering primary school produces its own organic meals. Children are encouraged to grow their own fruit and vegetables in the school's garden and this fresh, home-grown produce is then used as ingredients for the youngsters' school dinners.

presented to children they are very often very receptive, suggesting perhaps that the process of culture change is not as impossible as it first seems.

Other initiatives take our cultural relationship with food a step further by encouraging children to prepare meals themselves. We should never take good cooking for granted. It is one of the great skills and, like so many craft skills, it quietly teaches independence, enabling people to become liberated from the tyranny of the supposed convenience of the fast-food culture. For instance, recently there have been some marvellous popular efforts to highlight the benefits of cooking high-quality wholesome food. A recent renewed cultural focus on both growing and cooking food has in many countries helped to give life to a new Slow Food Movement. In fact the only thing about it that isn't slow is the speed with which it has spread – a clear sign that the values it embodies have struck a deep chord with the millions who desire nutrition that doesn't 'cost the Earth'.

Culture change is often one of the most intangible trends, but in the end it is also the most profound because it transforms how societies function. What I have tried to point out here is how economics and education are so vital for shaping those outcomes. And, in turn, both these things determine one of the most important trends of our age: population growth.

There is no doubt about it, monumentally controversial as the question may be, the problems posed by the predicted increase in the world's population cannot be ignored. The scale of the escalation is already astonishing. When I was born, in 1948, a city like Lagos, in Nigeria, had a population of just 300,000. Today, just over sixty years later, it is home to twenty million. Thirty-five thousand people live in every square mile of the city and its population increases by another 600,000 every year. The same eye-watering statistics apply to many other mega-cities around the world. Whichever one you choose, the population is increasing fast. Worldwide, the number of people goes up by the equivalent of the entire population of the United Kingdom every year, which means – and I know I have mentioned this earlier, but it is so important – that this poor planet of ours, already struggling to sustain 6.8 billion people, will somehow have to support over nine billion people within fifty years. To take a snapshot of the implications, take a look at the Arab world, where 60 per cent of the population is now under the age of 30. This means that, somehow, 100 million new jobs will have to be created in that region alone over the next ten to fifteen years.

I am afraid we have to face this problem and not accept it as an inevitable consequence of how we live. How can any of these burgeoning cities ever catch up with the expansion in their numbers that they are seeing? How can they

hope to provide adequate healthcare, education, transport, food and shelter to so many? And how can the Earth herself ever hope to sustain us all when the demands on her bounty worldwide are becoming already so intense? At the rate things are going, the answer to all of these questions is that they probably cannot. We cannot make the equation balance, unless we seriously address how we stabilize and even reduce the human population of the world.

There are those who argue that better economic conditions and access to education help to reduce the birthrate. When people enjoy the benefits of development they tend to have smaller families, or so the argument goes. There may be some evidence for this. I have long been fascinated by Muhammad Yunus's Grameen Bank in Bangladesh, which offers loans to the poorest communities. The bank is now 90 per cent owned by the rural poor of that country and, interestingly, I am told that where the loans are managed by the women of the community, not only do those communities thrive but there is a tendency for the birthrate to go down. The same effect has been noticed as a result of a women's education project my charities have been helping to run in Satara, in a drought-prone region of Maharashtra in India. But this is small fry. It does not promise to make the sort of difference that needs to be made. With mega-cities growing as they are, there is little chance of the economic situation improving for most people and they certainly will not gain access to the sort of education that will help. I am afraid the very big cause of high birthrates remains cultural and this, of course, raises some very difficult moral questions. We each

An Orangutan in Kalimantan, Borneo, Indonesia. These apes are in danger of extinction due to large-scale habitat loss. Changed business practices could help these incredible animals survive, as well as the tens of thousands of other species that share the forest with them.

carry the same responsibility and it surely has to be asked whether it is not time we came to a view that balances the traditional attitude to the sacred nature of life on the one hand with, on the other, those teachings within each of the sacred traditions that urge humankind to keep within the limits of Nature's benevolence and bounty. This is a question that could never have been imagined in the days when the sacred texts of each of the religions were first written down, but the world situation is now so very different that we may be entitled to enquire whether the leaders of those faiths, as well as the world's political leaders, might consider the plight of the Earth, the 'sustainer' upon whom we all depend and the sacredness of all her life. We are all of us 'Sons of the Earth' and perhaps the time has come to ponder what is the responsible thing to do in the present circumstances – to think very carefully how large our families should be?

A Good Idea?

I believe that the many positive initiatives taking shape today offer real hope that it is possible to set out with a new purpose. Although we stand on the brink of considerable ecological upheaval that, for some, may spell catastrophe, there are still opportunities to forge a new way forward that cherishes different priorities; one that values wholeness and the interrelated elements of Nature; one that seeks harmony and strives to minimize social and environmental stress. This could be a hugely rich period of opportunity – with new, more sympathetic technologies, business models, new cultural values, and dynamic transformations generating millions of new jobs.

This renaissance could take place during the twenty-first century – if we draw on the ancient knowledge of Nature, and on the timeless wisdom that today survives in isolated pockets, surrounded by the monoculture of ideas reared on four centuries of progressive mechanistic thinking and industrialization. Let me repeat once more that this is not to suggest that this emerging period must exclude the learning of the industrial age, or discard the technological capabilities that we have honed. But it will place our incredible capabilities in a new context, amid a new set of priorities that once more unite our humanity with the Earth and the natural world that sustains us. Perhaps if Mahatma Gandhi were still with us he would regard this as a *very* good idea, and indeed not only a route towards civilization, but a civilization that might last.

My work over many decades has brought me into contact with many manifestations of what I might call 'the problems' and, as I have set out in this and the previous chapter, also the potential solutions. Threats to endangered species have been met with laws to protect them. As natural habitats were cleared, so Nature reserves were established to save some of what was left. As we have begun to understand the incalculable value of Nature, so in a predictable manner we have sought to put a price on the vital services she provides. But it is my firm view that we now need to consider the next stage, because for all of the positive steps that have been taken, achieving a sustainable future will still depend upon a fundamental shift in our perspective.

The renaissance that is starting to unfold is a flower that needs nurturing and it is still a very delicate thing. It will in part be built on our technical knowledge of how the world works and how our own machines work best, but its economic and cultural foundations have to sit firmly in a proper philosophy – that is, a 'philosophia', or love of wisdom. In the final chapter I want to describe what I mean by this; to describe what lies at the heart of a new way of looking at our world.

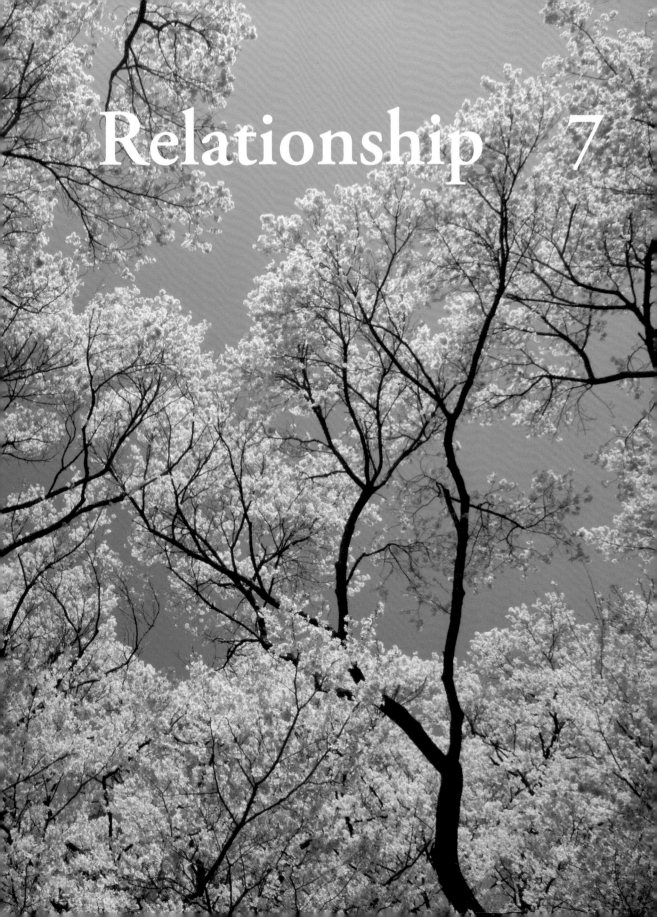

Relationship   7

There are more things in heaven and earth,
 Horatio,
Than are dreamt of in your philosophy.

WILLIAM SHAKESPEARE

I realize that in this book I have tried to take you on quite a journey, one in which we have explored the very shape of Nature's patterns and the central importance of her processes and cyclical economy. I have tried to chart what I believe is a fascinating journey from the ancients' understanding of harmony through to a present time when that wisdom has been blown apart by our rush for modernity. And I have done this because I wanted to show the fractured view of the world we have been left with. It is my conviction that the grammar of harmony, which was once to the fore in so many avenues of endeavour, from the simplest traditional crafts to the symbolism of the world's sacred traditions, has been cast aside and now lies at the margin, ignored, stamped out by an all-conquering preference for mechanistic thinking. I have wanted to demonstrate that the roots of so many of our problems lie in this decline. I am equally convinced that once people are allowed to become aware of the underlying principles I have to explain, many will find it a real revelation in terms of the true extent of our disconnection from Nature.

It can be a gloomy picture but I trust that the previous two chapters offered some rays of hope. They dealt very much with solutions – a variety of ways of doing things that work properly with Nature. Imagine the difference if these brilliant ideas were brought into the mainstream, rather than sitting as they do on the fringes and labelled 'alternative'. And therein lies the central problem, for to move on we have to overcome our cultural resistance to such ideas and the vision they represent.

This resistance has troubled me for a long time. I have often asked myself, what is it that maintains the collective reluctance to commit to what I called in my opening thoughts a 'Sustainability Revolution'? After all, we know very well what we are doing to the Earth. We also know that if we continue with 'business as usual' we may well drive it too close to the brink of catastrophe for us to find a way of pulling back. So what is it that prevents us from responding to the environmental crisis and allows us to go on tolerating so much destruction of our world and its incredible diversity of life?

PREVIOUS SPREAD:
Cherry blossoms

LEFT: *A bushman from the Khomani San community strikes a traditional pose in the Southern Kalahari desert, South Africa. One of the largest studies of African genetics published in April 2009 revealed how the San homeland could be the spot where modern humanity began.*

This is one of the most important questions of our day, it seems to me. Something stands in the way of 'right action' that might otherwise prevail over so much that is presently wrong. It is this obstruction I want to consider in this final chapter.

I believe part of the answer could lie in the importance we give to two terms: 'knowledge' and 'relationship'. The first is generally understood. We send our children to school to gain 'knowledge'. We send them on to university to acquire yet more. That knowledge of the world, we say, will equip them for life. And this may well be true. But what of our 'relationship' with the world? Is that something to which we give equal importance? I think not.

And yet the processes of Nature and her astoundingly beautiful patterns form our entire experience of the world. They are themselves the result of a rich biodiversity of life interacting with itself at many different levels. But for this finely tuned coherence to flourish it needs to be nourished and respected, treated with reverence and considered sacred. In our materialist world this may seem far-fetched, but we cannot afford any more to presume that the problems we have

I was absolutely shattered when I saw the devastation wrought by the Boxing Day 2004 tsunami. In this picture I am visiting the damaged Thiruchandoor Murugan Hindu temple in Navalady, Sri Lanka.

imposed upon Nature can be left for someone else to sort out. Certainly no solution will be found while we remain firmly glued to our television sets, transfixed by our PlayStations, or welded to our computer screens – no doubt being fed with yet more evidence of the extremely dangerous experiment we are conducting with the rest of life on Earth. To restore balance in the world, we must find the balance in *ourselves*.

I know all too well that suggesting we need to rethink and renew our relationship with Nature will induce howls of laughter from the usual suspects and unleash the full throttle of their mockery. But all ancient learning points to this need for balance within and without and I remain certain that this missing element of our *relationship* with the world contains the kernel of the cure to the many problems its absence has created. This will not be found in a scientist's laboratory, any more than it will be suggested by a government think-tank. This particular kernel of wisdom is to be found at the roots, where people who are closest to the Earth still live and have their being. Indeed, this book has aimed to show that unless we heed the warnings that come from the depths of human consciousness, where human nature is rooted to Nature herself, we shall unleash into the world uncontrollable chaos that no amount of clever techno-fixes will be able to deal with.

Against ever-increasing odds the so-called 'primary' peoples of the world still live in harmony with the Earth. Whenever I have had the privilege of speaking with them, no matter where they are in the world, they all talk of the importance of their relationship with the living Earth and almost always in exactly the same terms.

It is no accident that this wisdom is also to be found at the heart of the teachings of all of the world's sacred traditions. But there is a problem here. It is only too apparent that the specific ecological aspects of the teachings of many of the world's greatest religions have been forgotten in the Westernized world, or for some reason are no longer noticed. I am sure it is the legacy of the Age of Reason and then the European Enlightenment that has done this. Why should the religious traditions be any more immune to the corrosive scepticism of scientific rationalism than any other aspect of our tradition? Modernity has eroded the sensitivity of our religious relationship with the Earth as it has done everywhere else. Notice the subtle difference, for example, between the wording of the Lord's Prayer in the 1662 *Book of Common Prayer* and the way it is recited in most cases today in the Christian Church. The original line 'Thy will be done *in* Earth as it is in Heaven' has become today 'Thy will be done *on* Earth.' This might seem a small difference but the detachment I have tried to describe throughout this book is here too, and pretty much complete. So even in matters of religion we seem now to have reached a point where the absence of a proper understanding of our spiritual relationship with the natural world is all but comprehensive. Even the word 'spiritual' has been debased by the limited secular vision of our times. It is no longer understood as the unifying principle of Nature, the sense in us of the underlying core of the universe, that which impels the unfolding of what is, in truth, an endless moment of creation. We really have become completely numb – practically, ethically, and spiritually – to the many injuries we inflict upon the Earth. So let me explain what I have gained from my conversations with primary people about this word 'relationship'.

What to do when the great wave comes

Massive natural disasters always shock the world and wake us to what matters most, and none could have been bigger than the enormous tsunami that swept across the Indian Ocean on Boxing Day in 2004. At 7.58 that morning, deep beneath the Eastern Indian Ocean in an area known to geologists as the Sunda Subduction Zone, two of the Earth's rigid tectonic plates shifted, forcing the ocean floor to slide under the island of Sumatra. As the huge plates shuddered past each other, the strain of it generated a magnitude 9 earthquake that sent up a massive plume of energy through the column of water above. This great surge of energy radiated out across a wide area of the ocean until it reached the shallow inshore waters, where it rose into a wave of great height, reaching up to fifty feet high, before it crashed onto the land, flattening and swamping everything in its path over a vast area of South-East Asia. It even made its devastating presence known as far away as East Africa, some 2,800 miles to the west.

I saw for myself the aftermath of the wave and I will never forget what I witnessed. Whole towns were flattened, reduced to nothing but piles of giant matchsticks. The death toll was enormous. In those parts of Indonesia, Thailand and Sri Lanka where the wave did the worst damage, more than 200,000 people lost their lives. I was told it was the most destructive tsunami in recent history and the largest earthquake since the one that shook Alaska on Good Friday in 1964.

It was obvious that little could have been done to prevent such a wave causing destruction, although on my tour of one devastated region in Sri Lanka I learned that some of the impact could have been reduced had recent industrial developments in the region not damaged so much of the natural infrastructure. Along the Eastern coast of India, for example, miles of coastal mangrove forests had been cleared to make way for the rapid expansion of commercial shrimp farms – the very ones that also threaten the survival of those cone snails I mentioned earlier. Mangrove forests form a natural defence against the battering the ocean can inflict upon the coast, but they had become weakened and so the massive wave could do more damage. The effluent from those intensive farms had also polluted many miles of the coast, destroying stretches of coral reef that, to make matters even worse, had also been mined extensively. Again, these submarine barriers would once upon a time have helped reduce the power of a tsunami, but with such natural barriers so diminished, the wave was able to strike with full force.

I remember hearing the predictable response that we should consider how technology could be harnessed to create a sophisticated early-warning system so that people in vulnerable areas might be properly alerted the next time. I am heartened to hear that such a system has now been developed and tested and I pray that it will be effective when such a wave rises from the ocean again. But what fascinated me most on my visit was what was less talked about at the time, and still not mentioned much today: what happened far out to sea on that fateful day. It seems there was an early-warning system that was acted upon by those who could read it.

This became clear when rescue workers finally made it to the remote Andaman and Nicobar Islands, far out in the Indian Ocean. These isolated archipelagos, lying in the Bay of Bengal, are made up of more than 500 islands, but only about forty are inhabited, among them the Southern Andaman Islands, which lie about 340 miles to the North of Sumatra. As these islands lie so close to the very epicentre of the earthquake, the tsunami arrived there almost immediately, and yet many of their indigenous forest-dwelling people somehow managed to avoid its worst effects. It would seem that they did so because they knew the great wave was coming. They do not have electricity, nor do they

26 December 2004. People flee as a tsunami wave comes crashing ashore at Koh Raya, part of Thailand's territory in the Andaman Islands. This and a second wave tore apart the wooden buildings, with a third and largest wave coming forward and 'ripping apart the cement buildings like they were made of balsa wood'. Miraculously, all of the people in this picture survived.

The Mesjid Raya Mosque in Banda Aceh. The only building left standing after the tsunami of 2004, one of the deadliest earthquakes in recorded history. The Surin chain of islands in Thailand were spared this kind of devastation because the force of the tsunami was reduced by the coral reefs. Many other reefs around the Indian Ocean have been exploded with dynamite to help shipping and to create space for industrial shrimp farming.

have telephones or radios, so there is no system of sirens or flashing lights to warn them. They were told of the danger by Nature herself.

These people live a very primitive way of life, hunting and gathering, living in the forests and fishing from the shore. One community, for instance, is a tiny group called the Sentinelese. There are between 70 and 500 of them and they live on North Sentinel Island, which was completely uplifted by the tsunami, leaving the coral reefs that once fringed the coast high and dry above the sea. The Sentinelese are descended from tribes who first set foot on these islands tens of thousands of years ago and live in societies that archaeologists and anthropologists believe are similar to those that existed in prehistory, so they offer an example, rare these days, of how the earliest humans lived and organized themselves.

Rescue workers feared that this mysterious people would have been wiped out by the wave. However, much to their surprise, they found that the Sentinelese, together with the wildlife they depend upon, had survived relatively unscathed. In fact they seemed to have adapted to their new conditions with unexpected speed and flexibility. The islanders explained that they had done so because of their grandfathers' stories about what to do if they noticed certain changes. The Onge of South Andaman told of how afraid they had been when they saw the waters first recede from the shore but, like the Sentinelese, they had already noticed subtle changes in the behaviour of birds and even fish and, because such things are reported in their folklore, they responded immediately. They quickly moved to higher ground and the shelter of the forest and saved nearly all of their people.

This was not a unique response. This happened on many of the islands and it has happened before. I remember being in Northern Australia some forty years ago, where a terrible typhoon had devastated the city of Darwin, and I was told there, too, that the Aboriginal people, together with the birds, had sensed that a disaster was imminent and had disappeared to find a safer place twenty-four hours before the typhoon struck.

Reading the world

Whenever I have had the opportunity to speak with such people I have always been struck by their ability to 'read' the world in this way and how it is entirely dependent upon their relationship with the Earth. They talk explicitly about this relationship. To the aid workers who landed on North Sentinel the place seemed a tough, unforgiving jungle wilderness, yet to the indigenous islanders themselves their home is an intimately mapped garden. Their map is not written down, of course; they carry it with them. It is contained within their folklore and transmitted from generation to generation, in many cases in the form of stories. These tales are populated by the creatures that share their island and often appear as characters in mythical adventures, but the stories are peppered with advice and moral guidance. There are many references, for instance, to the medicinal properties of berries and plants, as well as lessons in social responsibility and manners of behaviour. In this way their folklore is far more than a form of entertainment to be enjoyed in the evening sitting around the fire. It emphasizes the moral code, it educates the young, it reminds everyone about how best to survive and, in the case of the Sentinelese, it tells them what happens when the big wave comes.

These people could not have responded to that imminent disaster had they not been acutely sensitive to the meaning conveyed by a body of folklore that teaches them to be aware of subtle changes to their surroundings. I happen to believe that this is not unique to primary people. It is common to us all and we, too, have a folklore tradition that teaches the same principles. It is the same sensitivity to change and the innate ability to anticipate peril, surely, that tugs at the gardener's conscience when he or she discovers Spring flowers blooming in Autumn, or the birdwatcher who records the first eggs of the breeding season appearing on earlier and earlier dates, or even the NASA scientist who studies the patterns of ice-melt throughout the Earth's history? Even the non-specialist will feel a twinge of consternation about what potentially lies ahead for humanity. There is a general undercurrent of concern, a nagging feeling that the modern approach goes against the grain – that somehow it is not right. The only difference between those of us in industrialized nations and those who still live in primary cultures is that the noise, fuss and fumes of our modern world drown out the sound of Nature's warning – to some extent, on purpose. We may well sense that problems abound, but our culture prevents us from responding in the way that the people of North Sentinel do – that is, immediately and without question – because we are all entrapped by the practical considerations of modern living and by the ties of modern economics, which have eroded our trust in those intuitive and natural feelings. The rational mind insists that we weigh all that we may feel against the economic consequences of responding as our instinct tells us to. And so we prevent ourselves from acting directly from the heart.

I find it revealing that a number of fire-fighting bodies both in the USA and Europe have decided in recent years to include exercises in their training programmes that encourage a response to the intuition, particularly in moments of great danger and stress. They have discovered that a purely mechanistic way of teaching procedures by rote, where dogmatic rules are drilled into fire-fighters, does not completely equip them to respond immediately when there is no time to think. Plenty of soldiers in the British Army have told me of life-and-death decisions that have to be made instantly when there is little or no time to analyze the situation rationally. They act 'with their feet on the ground', trusting their 'gut instinct' when something is not quite right about a pathway or the look of a building. In this way many a soldier has escaped a land mine. I am told that experiences such as these in Iraq have led the US Marines to study them seriously and to discover that it is possible to improve a soldier's intuitive response through training. It is something stressed when the US authorities train their Homeland Security staff. One of them explained that it was her intuitive sense that something was wrong that made her challenge the so-called Millennium Bomber when he tried to blow up Los Angeles Airport on New Year's Eve in 1999. And you only have to think of the many remarkable inventions, not to mention works of art, that have come about through someone going with their gut instinct. The same can happen in business with a snap decision that goes against all of the logical arguments learned in business school, but turns out to have been just the right thing to do given the circumstances of the moment. Why, then, is our intuitive response so derided? Is it

Service personnel and firefighters are trained to not only master technical methods but also respond to their intuition. All of us are equipped with an 'inner tutor', and by hearing what she says we can sometimes detect danger in ways that are not open to our rational minds. Perhaps if we listened to our intuition more we might embark on more sustainable and harmonious ways of living.

not time for us to re-sensitize ourselves to what Nature is telling us? Is it not time for us to become so immersed again in Nature and in her ways that we, too, are as alive to the disruption we cause to Nature's essential systems as those who live deep in the remotest parts of a rainforest or on tiny islands in the wide open sea? If it is time – and surely all of the warning signs tell us that it is – how might we do this?

Primary learning

If I were to boil down into one simple definition the philosophy of so many of the world's primary cultures, I would say that they know the world that knows them.

There is no doubt that our many astonishing advances in technology prove we know the world – how it works and how it can be mechanically manipulated – but any notion that the world itself is knowing and that we are part of that knowing is an idea that the Western outlook finds too difficult to digest. As I have sought to demonstrate, the modern attitude does not consider the Earth to be a knowing organism. We have come to see it as little more than a resource. So what might it mean to say that the world is also knowing?

In general, indigenous people still live very close to Nature and, whether they be the Sentinelese living on a tiny island in the middle of the ocean or the few remaining groups of Bushmen who roam the Kalahari Desert in the middle of one of the biggest continents in the world, they all live according to similar key principles.

Firstly, they are all very conscious of what I have already spoken of: that life is a web of interconnectedness; that it depends for its health and survival upon a complex interchange of mutual relationships, all of which are controlled by cycles of repeated patterns. They do not use such terms, but this is immediately apparent from the way they describe life, from their stories and from their craftwork. In their eyes, the Earth itself is a living participant in the system and ultimately a single living organism. 'She', not 'it', is the source of all nourish-

Kamayura Indian men and child, Xingu area, Brazil. Indigenous peoples often prove the most effective custodians of threatened ecosystems, such as the tropical rainforests. Their connection with land and Nature frames a perception of the world in which people are part of the living whole, not separate from it.

ment and as the Mother her nourishment goes beyond providing food. All that they know, all that they dream of in their imagination, all that they say, comes from the Earth. And this feeds a third common principle at work in all primary cultures: they have a profound spiritual reverence for the Earth. They refer to it by different names but they all see themselves in communion with what might be called 'the underlying intelligence of being', which bestows upon them the responsibility to safeguard the harmony and connectedness of all life. Through their role as stewards of the world they have a spiritual relationship with Nature. This means they inhabit the world from the inside out. They take care to maintain the inner harmony they perceive within themselves, which they contend means that the balance in things will be maintained in the outer, material realm. If they fail to do this, all of these age-old cultures predict, there will be discord and the Mother's sources of nourishment will dry up.

I find it fascinating that so many diverse First Nation people talk in this way about the world because it is precisely this model of the world that is presented by the ancient Greek philosophers and by the mythology and symbolism of the other great periods of civilization we dipped into earlier. In all cases, the essence of life is considered to be a mysterious, benign, sustaining force that seeks expression through its 'actualization' in the material world. This force lies at the core of life and it is perfectly natural that we should seek communion with it because we are very much a part of it.

Just as Plato did 2,400 years ago, or the later 'Neoplatonists' like Plotinus of Alexandria, primary people today see this core as 'shape-forming'. They consider everything they do in the world to be an event formed first in the spiritual realm, where everything has its source. Back in the third century Plotinus was very clear about this. He was in no doubt that consciousness gives rise to matter, not the other way around.

This is why so many First Nation people refer to reality as being a kind of dream state. The Kogi Indians of Colombia, for example, who live high in the Sierra Nevada de Santa Marta, call this state Aluna. Far away in the Kalahari Desert in Southern Africa the few remaining Bushmen hold to their belief that 'there is a dream and it is dreaming us', which has resonances in the Vedas of the Indian tradition, where the world is described as 'Vishnu's Dream'. And perhaps most famously, it is there in the outlook of the Aboriginals of Australia, who talk of all life as coming out of Altjeringa, which means 'the Dreamtime'. The Aboriginals consider that the Dreamtime gives the world its presence. It is this presence in life that 'knows' man. The land lives and, as humanity is a part of the land, so the land inhabits man. We are *in* Earth as we are *in* Heaven.

All indigenous cultures talk of listening to what the land says and many see life as operating in two forms of time – one that frames their daily activity and one that is higher, the infinite cycle of spiritual time, which is the Dreamtime that does the dreaming. The Australian Aboriginals say the Dreamtime is more real than outward reality and what happens in the Dreamtime establishes the values and the laws of society. Hence their particular warning, so chilling in today's world, that 'those who lose the dreaming are lost'.

Whatever we might make of this view of reality – and, of course, it is only too easy and fashionably cynical to dismiss all this as primitive superstition which has no place in today's modern, progressive, 'civilized' society – there is no denying that because of this perspective, which is, incidentally, not just thousands but tens of thousands of years old, not one of the remaining primary cultures of the world considers itself to be a master of creation. In such cultures that would be tantamount to blasphemy. The Kogi Indians of Colombia, for example, call themselves the servants of Serankua, the Father who created the Earth and the Sky but specifically made the Kogi people in order to care for all else in Nature. The Kogi refer to themselves as the Elder Brother, created by Serankua to protect the Earth, whom they inevitably call the Mother. In their system there is also a Younger Brother, a wayward creature who cares not for the Mother, who wilfully destroys what Serankua has created, and whose ways must be curbed before it is all too late. No prizes for guessing who that Younger Brother is. They say he was the one who went away across the ocean and has now returned to destroy every corner of the Earth, including the very tops of their most sacred mountains where the glaciers have melted, the tundra has died and the rivers and streams are drying up.

Such cultures are, of course, not without fault. Humans are complicated and imperfect creatures and there is always friction wherever societies form, but evidently such people the world over cannot conceive of themselves as being disconnected observers of the world. Just as a fish would find it pretty impossible to understand the concept of water, given there is nowhere in the vast ocean where it could somehow separate itself from the substance that sustains it, so people of primary cultures do not conceive of there being any distance between themselves and the rest of creation. Creation is a living presence, woven into the fabric of everything, and their communion with its presence offers knowledge of the world.

There are echoes of this sense of participation in what is now the developed, Western world – which was also once so alive to it. Many of the world's designated national parks are often sites that were once considered sacred by the ancestors of those who live there now. There are many mountains in China, for example, that have been held to be sacred by Daoists and Buddhists for more than 2,000 years. Today such areas form the core of some of our finest World Heritage Sites and national parks. Not without a fight, though. They have come under intense pressure to be part of agricultural and urban developments, but local people have felt so passionately that these sites should be preserved that they have struggled to protect them as special places embodying the core sacred values of their traditions.

Humans are complicated and imperfect creatures and there is always friction wherever societies form, but evidently such people the world over cannot conceive of themselves as being disconnected observers of the world.

I regard this impulse to ring-fence sacred spaces as a remnant of that 'golden thread' I described earlier. It glints with a dimmer glow amid the steel and concrete of our industry-filled landscapes and it calls out to be revived. I cannot help feeling it would be a good thing if this happened on a much wider scale – that we do the same for Nature as a whole, rather than restricted corners of particularly beautiful landscapes. What would be the outcome if we declared Nature as an endangered species or the whole of it a World Heritage Site? What if we gave Nature rights?

For us to even consider that such ideas might not be mad or bad but sound and of value we have to be ready for a pretty dramatic shift in our outlook. It would not be just a kind of intellectual shifting of opinion, as simple as changing a political or a commercial affiliation. If it only happens at that level, the world will carry on as before, having swallowed a dose of 'green wash' to make it feel better. To make a substantial long-term difference the shift has to be much more fundamental. It requires the guts to bring the two sides of our own nature back into balance through a comprehensive re-grounding of our awareness of that very 'core' of life I have just described. Only this will free us from the binding mentality of the current world view. So it is a task of quite literally 're-minding' ourselves and 're-membering' our culture – that is to say, putting all of the bits of our fractured psyche back together again so that we transform the way we think and feel about the world according to what lies deep within us all.

As we have seen, once upon a time this rooted approach was not just active in remote indigenous communities. Western civilization was equally rich in instructive myths. Just as with the teachings of spiritual texts, the lessons are all there, but we do not seem to notice them any more. Consider how Homer, in his great epic the *Iliad*, describes what the gods do. Even when he tells us of something as simple as the rain, notice that it is not Zeus making it rain, it is Zeus raining. Or when the Sun comes up, Homer's familiar description of Aurora's 'rosy fingers' is not mere poetic licence. The dawn *is* Aurora. The ancient Greek view, like those of all other ancient cultures, hinged on humanity's religious relationship with Nature. The myths that sprang from that relationship carry the same instructions about respecting Nature's limits. No end of heroes suffer a horrible fate when they offend the gods. The story of the hunter King, Actaeon, for instance, who offends Artemis, the goddess of hunting, and is turned into a stag and torn limb from limb by his own hounds. Or the boy, Phaeton, who brags that his father, Helios, is the Sun god and demands of him proof of this by letting him drive the chariot of the Sun across the sky. The chariot is too hot to hold, the horses too fiery, and the boy is consumed by the flames as the Sun careers out of control, burning the Earth, boiling the sea and disrupting the balance of the whole universe.

The eye of the heart

When those myths were young it was the bards who told the stories and today it is still the poets who carry the message, even if they are less heard than they were. I was immensely fortunate to know the late Kathleen Raine, one of the great British poets of the twentieth century, whose emphasis on the voice of Nature was so out of phase with the mood of her times. Her formidable spirit never gave in to the prevailing tide that grew ever greater during her lifetime. As she put it once, faith becomes the victim in 'a world that does not believe in Soul'. But as she said this she told me we should never give up the battle. She had carried the flag for spiritual understanding as a Cambridge scholar in the 1920s as Modernism grew into a powerful intellectual ideology. Rather than follow the bright new lights, she chose to clarify the understanding at work in William Blake's writing. Blake was hardly in print in those days, but her colossal scholarship not only helped to put him firmly on the map, it also demonstrated just how wide and deep his reading had been. Kathleen lived until her nineties and over her long life she charted in her poetry the decline of our sense of the

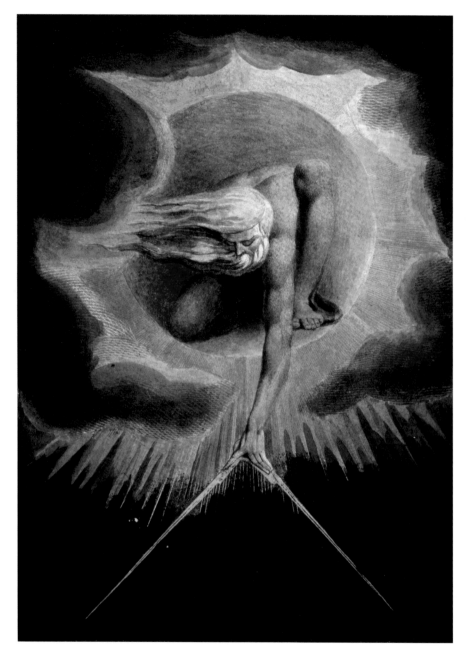

The Ancient of Days, by William Blake, UK, 1794. Urizen, depicted as a bearded old man sits above the world as the embodiment of conventional reason and law. Sometimes he bears architect's tools to create and constrain the universe, or nets with which he ensnares people in webs of law and conventional culture.

sacred and our detachment from what Nature speaks of. Towards the end of her days, when others with less energy might have considered slowing down, she founded the now much-admired Temenos Academy in London with the specific aim of helping to reverse the premises of what she felt was a spent, materialist civilization.

I am very proud to be the Academy's Patron because, central to its

endeavours is the study of what Kathleen Raine called 'the learning of the imagination'. As she put it in her poem 'Mandala', 'Wherever the eye falls, the mystery begins to unfold. It is there, the growing point of love, an ever-opening rose.' In her view, the imagination is our point of contact with 'the eye of the heart' – a term used by the great thirteenth-century Persian Sufi poet Jalaluddin Rumi, who often describes the imagination as 'al basira', which means 'insight'. It is the eye of the heart that penetrates the outward reality of all things to enable us to see their inward reality. William Wordsworth understood this well when he described the need of being attentive to the 'sense sublime … a motion and a spirit that impels all thinking things, all objects of all thought, and rolls *through* all things' so that 'with an eye made quiet by the power of harmony, and the deeper power of joy, we see into the life of things'. An admirer of Wordsworth who also became a revered poet in England, Alfred Tennyson, described his meditative experiences as transcendental and how he sensed the ineffable presence, the sacred breath of life, 'dissolving of the limits of selfhood until the infinite alone seems real'.

The Temenos Academy has also set out how this learning has shaped the culture of the Islamic tradition. A central principle taught in the universities that flowered from the eighth century onwards in the Muslim world was that things cannot be understood in isolation. Subjects were not taught separately as they are today. Instead, any one thing could only be known in a connected context within the universe. It is not by accident, but a direct result of this, that

Islamic art and architecture are so very mathematical. The distinction did not exist as it does today between academic subjects like art and mathematics. Mathematics was a kind of spiritual mathematics in effect and 'science' per se was seen simply as knowledge – a knowledge that was the intertwining of many fields of learning. This is why the feminine forms of Nature found in everything from a rolling hill to a natural beehive, from the rhythmic flowing of the waves to the patterns made by the growth of plants, are perfectly intertwined with the properties of the straight line and the angle – shapes that stand for what is fixed and stable and give the universe its order – to create beautifully balanced pieces of decorative art that display both a mathematical knowledge of the world and the deep-seated interconnectedness of all things. It is for this very reason that the Ka'bah, the focal point of the great Mosque in Mecca, is a cube, around which the thousands upon thousands of pilgrims circle during the annual Hajj. The geometry of the cube conveys the impression of clarity, order and calm and, as the pilgrims progress around the great black cube, so the Ka'bah itself stays still, because ultimate truth at the heart of life does not change. There may be many thousands of pilgrims, but there is one truth and it sits at the centre of things.

The implication of the unity that lies at the heart of life is just as apparent at the domestic level in traditional Islamic houses, which are often built around a central courtyard or garden. The light does not come from windows facing out onto the street, but pours in from the sky directly into this courtyard where, perhaps, a pool babbles in the middle with the waves from a gently bubbling fountain, so that the whole house is inward-looking and focussed on this tranquil and peaceful centre.

This is not 'romantic nonsense'. It is an outlook derived from a profound study of the curious interrelatedness of the world around us. In the Sufi tradition it is said that if it were possible to chop up the realm of time and space into an infinite number of pieces, the whole of the universe would still be present at every location. The Sufi poet Mahmud Shabstari puts it beautifully in 'The Mystic Rose Garden':

> Know the world is a mirror from head to foot,
> In every atom a hundred blazing suns.
> If you cleave the heart of one drop of water,
> A hundred pure oceans emerge from it.
> If you examine closely each grain of sand,
> In its qualities a drop of rain is like the Nile.
> A world dwells in the heart of a millet seed.

This has striking echoes of the famous view of reality taken by the English poet William Blake, who was, as Kathleen Raine so ably demonstrated in her impressive work of scholarship, *Blake and Tradition*, very well versed in the wisdom of Islam. Consider the famous opening lines of his 'Auguries of Innocence':

> To see a world in a grain of sand,
> And a heaven in a wild flower,
> Hold infinity in the palm of your hand,
> And eternity in an hour.

The point is clear in both examples. Everything is bound to all else; causes are linked to effects and mindful of a unity that underlies the apparent diversity in the world.

I quote these poets because they help us identify Wordsworth's 'sense sublime', which is just as important as knowing how natural processes work. We need modern science, for sure, because without it we would have a more limited knowledge of the world, but let us not forget what the sacred texts revealed, what poetry and the other arts give us, which is reverence. Reverence is not science-based. It is not knowledge. It is an experience induced by love, and love comes from relationship. Without reverence and love, without a spiritual relationship, it seems to me that we are little more than a chance group of isolated, self-obsessed individuals, unmoved by love and un-anchored by any sense of duty to the thing that deserves our reverence. And so we fall into the trap of thinking that we are free to act without responsibility, legitimized by the undisputed right to do what we want to do, which, as I have tried to show, so often does great harm.

The lesson for me from all of this is that if we fail to reinstate a much deeper awareness of how the world we inhabit really works, as traditional societies quite clearly do, we must expect an even wider disconnection, both from the Earth and within ourselves. The two are intimately connected. The destruction of Nature is ultimately the destruction of our own inner being and it is this inner destruction that, if not checked, will lead to yet more destruction of vast numbers of species of animals and plants. It is a vicious circle that grinds away at human well-being. I believe that we can halt this course of events if we properly recognize the difference between a world based purely on knowledge and one that balances this knowledge with what we gain from our relationship with the Earth.

Being a passionate gardener I know all too well that it is important to know

how to garden – when and where to plant, how and when to prune, and so on. However, my knowledge of gardening, such as it is, is not my *relationship* with my garden. That comes from my experience of it. I employ my knowledge to create the experience that enriches my relationship and the meaning I gain from an understanding of the natural world. Having enriched and sensitized my relationship, the plans I may make and then execute will be based upon enriching my experience of my garden yet more. A right and beneficial relationship is the purpose of the process. The exercise of knowledge is merely the means.

Consider that distinction in something like factory farming, which is so dependent upon a detailed knowledge of how the natural world behaves and yet any *relationship* we might have with the processes involved has been completely and deliberately obstructed. All of the knowledge is applied behind very high walls and tightly guarded gates that allow no relationship at all. In this way it is fair to say that the industrialized approach in its most extreme form could be called 'knowledge without relationship' and this is why for so many years I have been at pains to point out why we cannot simply rely upon

The stump garden at my Highgrove home in Gloucestershire. This is a large space, but through understanding local conditions it is just as possible to nurture the most beautiful garden in even a small space. I am constantly impressed at the creativity that lies behind the many wonderful gardens that I have seen. They are direct reflections of people's connection with Nature.

yet more technology – our knowledge – to solve the many problems we face. We have to find a way of reintegrating a proper sense of our spiritual relationship into the mainstream mix for there to be a healthy, balanced approach. Without the demands of a relationship we can be as extreme in our use of knowledge as we like. So, in our flawed dialogue with the world, the fault does not lie in the absence of knowledge. It lies in the absence of any relationship.

Playing a part

I can say this with some confidence because I have seen that when this wider perspective prevails it can make such a big difference. There are now a number of examples around the world where indigenous, primary societies have decided to take a stand against developments encroaching on their way of life and their environment. There are also some inspiring examples where the sacred traditions have rediscovered the teachings in scripture and applied the ideas with encouraging results.

In the Amazon region, for example, where a large arc of dense rainforest has steadily been eaten away by developers across South America – releasing, I might add, more carbon dioxide than all of the cars and planes in the world put together – dozens of initiatives have been launched to try to stem the rapid clearance. However, one solution above all others has been effective, and that is the transfer of control to indigenous people.

Indigenous people demonstrate against the Belo Monte Hydroelectric Power Station in Brasilia, Brazil. Set to be built in the Brazilian state of Para, this is one of the main projects within the Brazilian government's Acceleration Growth Program. People from the Xingu reserve are protesting against the prospect of their land being flooded.

Manhattan skyscrapers, New York City, USA. As the world population becomes more urban, our collective connections with Nature have been severed. With more and more people living in cities, the trend is set to continue.

Maps showing recent patterns of deforestation across this vast region demonstrate how dramatic the difference has been. In many of the worst-hit areas, where development has happened outside the indigenous reserves, the process of deforestation is all but complete, but where the indigenous people have gained control of the land, the forests have largely remained and both the emission of greenhouse gases and the loss of biodiversity have slowed down. This has happened not simply because the Amazonian people have been granted legal entitlement. It has far more to do with the freedom they now have to act according to the way they regard the forests. The maps that portray this difference not only show where the forests remain, they also demonstrate the fundamental difference between the outlook of those who wish to clear the forests and the philosophical perspective and intuitive feeling for the Earth of the indigenous people who are its guardians.

Whereas the national land-use planners, the energy firms, the agricultural commodity companies, the loggers and the miners all see the forest as a source of financial revenue and a resource-rich opportunity for fast economic 'growth', the indigenous communities see the forests as their home. It is their sustainable livelihood. They regard it as their irreplaceable heritage that must be passed on intact. They stress this when they educate their children, so why couldn't mainstream education systems try something similar? After all, there is now plenty of evidence from around the globe that when traditional principles are reintroduced, people change their behaviour for the better.

The Islamic Foundation for Ecology and Environmental Sciences (IFEES) has been successfully demonstrating this to be true in another part of the world. It is an organization committed to engaging Muslims in the steward-ship of ecology. They recognized what I hinted at earlier, that the sense of stewardship that once endured within their tradition had suffered and become weaker, mainly because of the influx of the sort of commercial, economic developments that come with progressive forms of urbanization. Some years ago they began working with the teachings of the Qur'an, going directly into communities and engaging with local leaders to bring about change from the bottom up.

In Zanzibar in the Indian Ocean, for example, they have been working in fishing communities where, for years, people had been using dynamite to stun coral reef fish to make them easier to catch. Conservation groups had sought to discourage the practice using rational arguments, suggesting that blowing up the reef is a self-defeating method. Dynamite may stun the fish, but it also destroys their habitat, leaving the next generation nowhere to be nurtured. While this strategy had limited success, the work of the IFEES has been much more effective. Shortly after it ran local training seminars, there was a dramatic reduction in the amount of fishing with explosives. Rather than teaching a rational argument, it found success by reconnecting fishermen with the basic principle of stewardship that is found throughout the Qur'an.

It was the eminent Islamic scholar Abdullah Yusef (who brought such a deep understanding of the Qur'an to the West in the 1930s) who put it so well in his commentary on the Suras when he described humanity as being given 'spiritual insight so that we should understand Nature, know God through His wondrous signs and experience the sublime joy of being in harmony with the infinite'.

   This impulse also lies at the heart of the work done in the Mandailing region of Northern Sumatra, in an area surrounding a national park where illegal logging had been rife and settlements were beginning to encroach on the precious forest that was left. By reminding people of the timeless wisdom of Islam in a programme that began by reaching out to local faith leaders and then to the people at large, the community was encouraged to work according to Islamic ideals that stressed the importance of the management and care of natural resources. The stress, once again, was not so much on our knowledge of how the world works, but on our deeper relationship with the rest of Nature that sustains us. As Fazlun Khalid, the founder of the IFEES, has said, 'It is as

Return of fishermen in traditional dhow boats, Zanzibar, Tanzania. Communities reliant on natural resources are sometimes more influenced by spiritual arguments than purely rational ones in building more sustainable livelihoods.

if a shell was punctured when this work took place. The knowledge lay locked inside, but when the shell was broken the knowledge and practical wisdom freely gushed out.'

We do all have a feeling for the natural world. I am greatly moved by the many examples I have seen or heard about involving young people, perhaps considered disruptive at school or worse, who have been transformed by their sudden contact with the natural world, some of which I mentioned in the previous chapter. There are some excellent examples of farms that give city children from difficult backgrounds hands-on experience of being in Nature – some particularly fine projects around the UK do this by putting young people excluded from school or college in charge of horses – and the transformation that this brings about is so dramatic I really wonder why more schools and colleges do not adopt the same approach for children who are not so disruptive. The experience is clearly beneficial and often life-changing. So why not reintroduce school farms and gardens, with some livestock whenever suitable? This would be by far the best way to reconnect children with Nature and with their inner Nature.

It was the biologist Edward Wilson who coined the term 'biophilia' to describe our empathy with other forms of life. From his research, it seems that we do have an urge to affiliate with Nature and we benefit when we do so. Consider how many people are drawn to walking the hills, sailing on lakes, wandering on moorland, or experiencing other kinds of wilderness. The same urge to reconnect is also visible in many everyday activities and pastimes that defy rational explanation. Some people who lead busy lives have always managed to find the time to grow fruit and vegetables. The benefit may well be that they can eat their produce, but when you consider the amount of time and effort involved, these vegetables are very costly to produce, and yet now more and more are doing so. Recently there has been such a rise in demand in the UK for small plots of land to grow food that in some towns the waiting list is now years long – and this, after so many allotments have been sold off by local authorities for development.

However, in these small ways the tide may be turning. I was greatly inspired by the way in which a more rounded view of Charles Darwin's work seemed to emerge during the recent celebrations to mark the bicentenary of his birth and the 150th anniversary of the publication of his book *On the Origin of Species*. He had often been associated with establishing the modern atheistic view that Nature is nothing but a Godless battleground of competing species, but this is not necessarily how Darwin himself saw things, even if it is definitely the way the present-day 'neo-Darwinists' like to portray him. In his own writing he

does not seem to describe quite the allegiance to a mechanistic model of evolution that they do. His words can be surprising. Take this reflection, noted in his journal in October 1836 during his journeys in the *Beagle*, a critical period in the formulation of his ideas on natural selection and evolution: 'Among the scenes which are deeply impressed on my mind none exceeds in sublimity the primeval forests un-defaced by the hands of Man; whether those of Brazil, where the powers of life are predominant, or those of Tierra del Fuego, where Death and Decay predominate. Both are temples filled with the varied productions of the God of Nature: no one can stand in these solitudes unmoved, and not feel there is more in Man than the mere breath of his body.' Although Darwin was probably the greatest scientist of his age, there is little doubt that the indigenous people of the Andaman Islands would know exactly what he meant.

I often think of Darwin's studies of the natural world when I hear that phrase 'the survival of the fittest' being used most inappropriately in the world of economics and business. It has been hijacked specifically to justify the ruthless 'me first' outlook that now abounds in so much of the commercial world, where it is perfectly reasonable to act according to the brutal, dog-eat-dog 'law of the jungle'. This is despite the fact that, as Darwin would have pointed out, the jungle metaphor is just as inappropriate. A jungle comprises many communities that are interrelated and interdependent. Yes, there is competition but there is also a vibrant interaction at play that enables energy and nutrition to be

On a visit to the Galapagos Islands I met this giant tortoise. I was told that Charles Darwin himself met this majestic creature when he visited the islands in the 1830s, during his world-changing voyage on the Beagle.

Aeriel view of a river valley full of silt from erosion, Madagascar. Among the ultimate casualties of our behaviour and our approach to how we treat the Earth will increasingly be humankind. Major threats to food security could emerge during the 21st century, in part because of soil loss in turn caused by deforestation, over-grazing and unsustainable farming.

processed and circulated so that the entire forest functions healthily as a whole. Words matter and the way we use them matters too.

So, what would the difference be, I wonder, if we applied a more proportionate view of Darwin's understanding of Nature; one that brought to the fore the true meaning of 'the law of the jungle'? It would lead us, surely, to appreciate the vital importance of the sort of relationships I have been seeking to define – with each other as well as with the rest of the natural world. Perhaps we would start to see organizations as organisms that work to restore and safeguard diversity in our economy; to reward collaboration and interdependence; to build skills that nurture complexity rather than obliterate it with monocultures; to stress the use of materials that can be recycled so as to eliminate waste and to maintain all of the subtle checks and balances that could keep an economy, as an ecosystem, vibrant and healthy. It very much depends upon the way we look at the world.

It is one of my big regrets in life that I never met Lady Eve Balfour, who was such a courageous pioneer of a return to organic, sustainable forms of farming in the UK. In many ways she was someone after my own heart who was quite happy to make certain people apoplectic with her calls for farming to turn once more to traditional practices. Many in the agricultural business still think that this means an end to 'progress' but, as I have tried to show throughout this book, there is great wisdom in respecting Nature's limits and accepting restrictions in the name of sustainable husbandry. Eve Balfour founded the UK's Soil Association in 1946, after her book *The Living Soil* had met with great success three years before. Way back then she could already see the disastrous consequences of modern farming systems that now think nothing of treating animals as machines and the land as if it were an industrial process in a factory or, as I have suggested here, as an extension to a laboratory. In 1977 she reflected on those early days, observing that her fellow organic pioneers 'all succeeded in breaking away from the narrow confines of the preconceived ideas that dominated the scientific thinking of their day by looking at the living world from a new perspective – they also asked new questions. Instead of the contemporary obsession with disease and its causes, they set out to discover the causes of health. This led, inevitably, to an awareness of wholeness and to a gradual understanding that all life is one.'

This statement bears a great deal of contemplation. Once again it concerns itself with the relationship between all things and the balance that is the *cause* of health. It could also quite easily have been made by any number of the figures I have made reference to in the previous chapters. 'That all life is one' was certainly a central tenet of the alchemical philosophy of ancient Egypt, as it was a central principle of the Neoplatonists in third-century Alexandria. It was the view of Dr Dee and his circle of Platonists in Tudor England, as it has always been the view within Sufism. But here is someone putting the philosophy into practical action and I would urge this to happen in many other fields of enterprise too. We need far more of this joined-up way of thinking based, as it is, upon the health of things rather than upon their exploitation.

Harmonic thinking is found to be effective not just when applied in organic farming, but also in the kind of thinking that produces town planning that feels right and conjures communal well-being. It is certainly there in traditional house-building, boat-building and many other tried-and-tested craft skills. So could it be applied to those areas where it does not exist: in the way we design and manage our transport systems, in hospital design, in the diagnostics and treatments in healthcare through proven disciplines like osteopathy and acupuncture, and through a much more joined-up way of providing education? Could the approach taken by my School of Traditional Arts mentioned earlier be taken to a higher level, with the principles it teaches being demonstrated not just in week-long workshops in schools in one town, but as part of the curriculum of teacher-training colleges?

It was Mahatma Gandhi who pointed out that 'the difference between what we do and what we are capable of doing would suffice to solve most of the world's problems'. How true that is. It is not so much a matter of capacity, more of deciding to do something. And we might very well begin by embarking on a wholesale reappraisal of all that we actually know, for all that we need to know is already known. The starting point is to see things differently; to shift our perception from the current, dominant world view that fills the spiritual vacuum with yet more material consumption, neglects our responsibilities through the excuse of technology and widens the social cracks with wedges of a selfish individualism. In so many ways, this approach is no longer relevant to the increasingly critical and completely different situation in which we find ourselves – it is no longer fit for purpose.

This is why it is of such profound importance that we understand we are not what we think we are. We are not the masters of creation. No matter how sophisticated our technology has become, the simple fact is that we are not separate from Nature. Just like everything else, we *are* Nature. Recognizing this fundamental fact should help us to adopt a much more coherent approach that may begin to shift our outlook from one that is reductive and mechanistic to one that is more balanced and much more integrated with Nature's complexity. Such an approach would recognize not just the build-up of financial capital, but the equal importance of what we already have: environmental capital and, crucially, what I have called here 'community capital'.

Harmony

It is my hope that with Tony Juniper's and Ian Skelly's help I have managed to demonstrate that there is much to be gained from the observance of the natural order and the rhythm in things, whether it be in the lines and shapes of architecture or the processes involved in agriculture, and certainly in the natural world as a whole. Not just because of the aesthetic experience this may bring, but also because it reveals how the same rhythms and patterns underlie all these things. Through the contemplation of the rhythms of life, it is possible to understand the forces that dominate everything we are aware of and to sense and gain from the harmony that exists between all things in their natural state. As all sacred traditions have sought to show, and as this book has attempted to demonstrate, the closer we dance to the rhythms and patterns that lie within us, the closer we get to acting in what is the right way; closer to the good in

RIGHT: *It seems to me that we have no choice but to live in harmony. By taking Nature as our tutor it might be easier than we sometimes think to build a more durable and more pleasant society. And what is the alternative? Carrying on as we are now, we know will court disaster. There is no option but to seek change, the question is, what kind, and how?*

life, to what is true and what is beautiful – rather than swirling around without an anchor, lost 'out there' in the wilderness of a view shaped solely by four hundred years of emphasis on mechanistic thinking and the output of our industrialized processes.

Studying the properties of harmony and understanding more clearly how it works at all levels of creation reveals a crucial, timeless principle: that no one part can grow well and true without it relating to – and being in accordance with – the well-being of the whole. We need to remind ourselves of this vital 'eternal law' again and again, it seems to me, so as to 're-mind' the world, using it as the gauge we apply to all we do.

Thousands of years ago our ancestors embarked upon a journey to derive comfort and security so that they could expand their knowledge of the world and develop ever more sophisticated tools with which they could use more effectively the natural resources they found in Nature. The progress they made was driven by immense hardships. The long march of investigation and innovation, the application of creativity and technology, the process that we now call 'progress', has procured for us immense benefits and opportunities, but it has also, paradoxically, brought us back to the place where we started from: facing a grave threat, surrounded by uncertainty and insecurity. The difference now is that we face those threats not as individuals and families, but as an entire global community, alongside much of the rest of life on Earth.

Unless we temper the side of our nature that has delivered such vast benefits, it could easily be the cause of our ultimate demise. It is already becoming the cause of the decline and looming extinction of countless species which, did we but know it, are part of the intricate web of life on which we all depend. In an age of rights do these species not have some sort of right to inhabit their particular niche? Do we not have some sort of responsibility towards them? It seems to me that we have to find ways to unite within our modern culture the perennial wisdom we have abandoned – the voice of our intuition that we have increasingly ignored as well as the spiritual essence of our being that now lies buried beneath great mountains of materialism. We need to understand that we are born into a universe that has meaning and purpose. All of the sacred traditions tell us that this 'purpose' is for life to know itself – this is the meaning of communion, by which we sense and help maintain life's essential balance.

This all ultimately depends upon how we perceive the world and our place within it. And this will mean somehow replacing our obsession with pursuing unlimited growth and competition with a quest for well-being and cooperation. It will mean shaping our culture so that its aims are rooted in relationship and focussed on fulfilment rather than on ever more consumption. If we can rebalance our perception and restore a sense of proportion to how we relate to the world – and on what basis we value the miracle of its marvels – it seems possible to me that we could create the conditions that ensure human societies thrive indefinitely.

There will be cynical critics who will scoff at such a utopian suggestion, but if they choose to dismiss such a vision, then it is incumbent upon them to come up with something better. One fact is clear. What seems to be their preferred choice, the one that prevails at the moment, which is to carry on as we have been doing, is not an option. An ever more divided and ecologically bankrupt world will be the consequence of our continuing in that vein and if we have any sense of responsibility for the future we cannot allow that to happen. We must recognize that to continue as we have been doing will only compound the problem.

The better, if not the only, effective course we can take is to see that we are part of the Natural order rather than isolated from it, and to appreciate Nature as a profoundly beautiful world of complexity. This world operates according to an organic grammar of harmony and is informed by the awareness of its own being, making Nature anchored by consciousness. In this way of understanding, life is seen as an interconnected, interdependent function of creation.

We do have within our societies and within our existing technologies the solutions that will enable us to transcend our current predicament. All we lack is the will to establish a more entire and connected perspective that includes giving space in our culture for the sense of the sacred in life; for reverence and even, dare I say it, for a touch of enlightened deference to Nature. After all, she *is* our ultimate 'sustainer'. Without such an integrated spiritual outlook, the many indications are that we will continue to deal with each individual crisis in a separate way, never seeing the connections that exist between them and the relationship we have with each element and the whole. And the consequence of that is a collapse of catastrophic proportions. Thus we stand at an historic moment. We face a future where there is a real prospect that if we fail the Earth, we fail humanity.

To avoid such an outcome, which will comprehensively destroy our children's future or even our own, we must make choices now that carry monumental implications. It is beholden upon each and every one of us to help redress the balance that has been so shaken by re-founding our outlook in a firmer set of values that are framed by a clearer, spiritually intact philosophy of life. Only then can we hope to establish a far more sustainable economic system; only then can we live by more rooted values; and only then might we tread more lightly upon this Earth, the miracle of creation that it is our privilege to call 'home'.

Acknowledgements

A great many people generously assisted the authors in the production of *Harmony*. They variously gave time to speak with us, to offer advice, to review extracts and to point us toward source material and images. We would like to thank: Mosaraf Ali of the Integrated Medical Centre; Robin Maynard at the Soil Association; Andrew Parker of Oxford University; Physicians Hugh Montgomery, Robin Stott, Michael Dixon and Michael Dooley; Sonia Roschnik of the National Health Service Sustainable Development Unit; Jules Pretty at the University of Essex; Pavan Sukhdev, Special Advisor to the UN Environment Programme; Liz Hosken at the Gaia Foundation; David Lorimer of the Scientific and Medical Network; Hank Dittmar, Ben Bolgar and Joey Tabone at The Prince's Foundation for the Built Environment; Dr Khaled Azzam, Professor Keith Critchlow, Paul Marchant, Dr Lisa DeLong and Delfina Bottesini at the Prince's School of Traditional Arts; Emily Shuckburgh, Robert Mulvaney, David Vaughan, John King, Corrine le Quéré and Eric Wolff at the British Antarctic Survey; Guy Thompson at Natural England; Euan Dunn at the RSPB; Andrew Simms and Ruth Potts at the New Economics Foundation; Dr Joseph Milne at the University of Kent; John Martineau at Little Wooden Books; Rupert Sheldrake; Jeremy James; Professor David Cadman; Vandana Shiva; Stuart and Julie Sender at Balcony Films; Tim Jackson at the University of Surrey; David Wilson at Duchy Home Farm; Simon Conibear at Poundbury; Justin Mundy, Jack Gibbs, David Edwards, Andreanne Grimard and Charlotte Cawthorne at The Prince's Charities International Sustainability Unit; Eric Schlosser.

Several people at HarperCollins also helped us put the book together, including Myles Archibald, Lisa Sharkey, Matthew Harper and Jonathan Burnham. We are also immensely grateful to Sióbhan Brooks and Catie Bland, Personal Assistants to HRH The Prince of Wales, who skilfully managed the comments as the various drafts we produced were circulated, always with calm and effortless professionalism. And among the many people at Clarence House, to Sir Michael Peat, Dr Manon Williams, Mercedes Luis Fuentes, Leslie Ferrar, Paddy Harverson, Patrick Harrison and their team who all helped manage the project, along with Robert Higdon and his staff at The Prince's Foundation in Washington.

Tony would like to thank his wife, Sue Sparkes, and their children, Maddie, Nye and Sam, for their understanding and assistance during the very busy period in which this book was put together, and it is with eternal gratitude that Ian would like to thank his wife, Juliet, and children, Emma and Sebastian, for their incredible patience and unending belief and support.

Index

Picture Credits